The Babel Factor

Other works
Banking on Babel: Survival & Profit
Through Language (2006)
by Robert D. Burgener

The Babel Factor

How Language Defines Difference

Essays on the role of foreign language

Robert D. Burgener, BA. MA

Barbara J. Murphy, MD

The Babel Factor

Library of Congress Control Number 1-91334481

"If you talk to a man in a language he understands, that goes to his head. If you talk to him in his language, that goes to his heart."

Nelson Mandela, Nobel Peace Prize winner

Contents

Foreword

The nature of our surroundings dictate differences will exist, if not be the norm. With so many variables and possible combinations, the outcomes are limitless and so we find or create comfort zones. We may define our personal comfort zone in terms of what we already know will happen with certain elements (objects or people) creating a set of boundaries that define this mix. When we encounter something we don't recognize, something different, we can be challenged or frightened by not knowing what will happen; or more to the point, how we will handle it. Do we fight or flee? Or as an old friend from Nebraska puts it, "do we get the bear or does the bear get us?"

This book offers a collection of observations on the theories of individual identity viewed from the perspective of language and the very real impact the inability to communicate in a timely, accurate manner has on the lives of people.

If the absence of difference is sameness, are there really any advantages in having every experience be an exact clone of what came before? The answer depends on who you ask. In Chapter 3 we will consider the pioneering work by William Thomas, and Florian Znanieck in a 1918 study titled The Polish peasant in Europe and America. They introduce us to the *Bohemian, Philistine* and *Creative* personality in terms of how each deals with differences. Somewhere along the continuum defined by no-change on one end and constant change on the other is a place where most of us live out our lives. From time to time our comfort zone is challenged and we may be asked or forced to confront our choices. We may decide that making a different choice is appropriate for whatever reason, or we may insist that we will not change our definition of what makes us comfortable just to accommodate the latest fashion of political correctness. The notion that we as individuals have choices is not universally shared in all cultures or political systems. In fact, if we want to understand how language defines differences we may need to look beyond political theories and social contracts because they tend to describe how issues such as language are dealt with by groups through laws and regulations.

The concept that we are all created equal before God or the law seems to defy the physical reality that we are not – at least where other human beings are involved. The beauty contained in a simple act of kindness by an individual is quickly overwhelmed by the ugliness of a mob. It is the memory of those individual acts that give us hope.

Political science and psychology provide analytical tools for understanding the delicate, dynamic balance between the needs of individuals with their differences and groups of those individuals who have created an alliance around some shared sameness. In his 1941 address to the U.S. Congress, President Franklin Delano Roosevelt described four freedoms which should be the right of all human beings:

"The first is freedom of speech and expression -- everywhere in the world.

The second is freedom of every person to worship God in his own way-- everywhere in the world.

The third is freedom from want, which, translated into world terms, means economic understandings which will secure to every nation a healthy peacetime life for its inhabitants -- everywhere in the world.

The fourth is freedom from fear, which, translated into world terms, means a world-wide reduction of armaments to such a point and in such a thorough fashion that no nation will be in a position to commit an act of physical aggression against any neighbor --anywhere in the world."

Desire for the freedoms articulated by one American president may seem basic to all human beings. In fact the basic themes can be found in religion and political philosophy in both the East and West from ancient times. Adding the words *everywhere or anywhere in the world* however, introduces the political dimension because while all parts of the planet are populated by homo sapiens with the same internal organs packaged in different shapes and colors, we as individuals have organized ourselves on another level as groups defined by geography, language and political affiliation. Erikson described this phenomenon as *pseudospeciation* in which groups seek to establish their own superiority based on some unique claim to human identity not shared by others. (Erikson 1996) As an extreme form of ethnocentrism, pseudospeciation goes beyond pride in the accomplishments of one's own group to deny the capacity for such contributions by any other group. Ethnocentrism itself creates blinders inhibiting our perception of others. As U.S. Marine Major General Douglas Stone, former Deputy Commanding General for Detainee Operations of Multi-National Forces-Iraq (2007-2008), pointed out in a speech at the U.S. Institute of Peace, ethnocentrism reduces the effectiveness of many Americans when they are abroad. "everything is not going to be the way it was back home." Americans are not the only group that behave this way. General Stone's capacity to appreciate other ways of doing things without depreciating his identity as an American officer provides one of those glimmers of hope mentioned earlier.

When leaders seek to define basic human needs in a context that supports political agendas, the results can be ethnic cleansing, for example, where the needs of one group require the elimination of another. Making the leap from promoting human freedoms to unilaterally imposing a form of government has not proven effective. In his book *Killing in the Name of Identity* Vamik Volkan observes: "Despite attempts at democratization in massively traumatized places such as Iraq, having a democratic state in so-called underdeveloped societies – or in so-called developed states for that matter – is no guarantee that conflict will not continue to be "solved" by war or violent means." (Volkan,2006:17)

Moving from world politics to the psychology of individuals, between 1943and 1954, Abraham Maslow developed his theory on the Hierarchy of Needs (Maslow, Motivation and Personality 1954). It seems there are several areas of common ground between these two examples of how we define human needs.

If we consider communication as a means of insuring not only our survival but the quality of our lives, then the words and gestures go much deeper than semantics or Hermeneutics. The point is that our experience with languages that are foreign to us –with differences- is not unique to any particular culture, age or socioeconomic position. It is part of the human condition.

People misunderstand each other when speaking the same language so there is clearly more to the process of human interaction than just our choice of verbal symbols. In fact, once we leave the comfort zone of our birth language, the choice is not entirely ours. We must either learn a foreign language or depend on the kindness of strangers or the legal terms of a service contract. We will discuss the issue of trust as dependency both in the dynamics of crossing borders between people and communities defined by differences in language, and as a reflection of our personal sense of vulnerability.

Understanding the role of language then requires at least a brief investigation of some of the theories of human behavior and our motivation, conscious and unconscious, for interaction with others. When the context of the interaction is defined by war or civil conflict and the people involved cannot interact verbally because they don't understand each other's language: then those choices of verbal symbols take on a particular significance. If the only phrases you've learned in another language center around concepts like "Stop or I'll shoot" or "Where's ali-baba?" you may be engaging in verbal exchanges that can save your life or the lives of others around you, but what impression do you leave with locals who are not a threat? The emotionally dirty business of sorting out who is a threat and who isn't has become increasingly difficult on the asymmetrical battlefield of today. The enemy no longer offers the courtesy of wearing a different uniform and the old designation of COB (Civilian on the Battlefield) has been replaced with SIN (Soldier in the Neighborhood.)

The choices, however, remain grounded in often unconscious concerns for survival and identity which offer themselves in a variety of ways in the examples presented in this collection of essays. Our focus is on the specific, observable events involving language and not on the psychological motives of the participants because the nature of a literature search which was the method used to gather most of the stories presented here, restricts comments to what has been reported. There was in most cases insufficient detail from which to accurately construct individual motives. The role of language in access to health care services by a particular group defined as refugees, for example, is commingled with larger issues of the national language of host countries and the multiple groups of economic and political immigrants competing for their own identity and share of the available resources. As such the issue of language cannot be treated in isolation from other behaviors, although we will see certain common denominators, such as the choice of interpreters, as a source of miscommunication for all groups.

The way we experience differences is deeply rooted in how we feel about our own capacity to survive change. If we are threatened by change because we doubt our capacity to adapt or adjust to a new set of circumstances, then we struggle to surround ourselves with familiar, predictable things and avoid or reject what is different or unpredictable.

When people come into contact with others who speak a different language and observe different customs pertaining to food, dress or worship there is a tendency to treat the differences as a novelty or curiosity. That works for tourist or exchange students who have voluntarily put themselves in a particular situation for specified period of time, but what happens to people when they are forced into situations where the time frame is less well defined or unknown?

Much of the emotion surrounding the subject of national language and those who enter a geographic territory as refugees or migrant workers speaking something other than the language of the geographic majority may be explained by a fear of the unknown and a lack of confidence in our own capacity to survive change. That is not an excuse for arrogance, bigotry or just bad manners on the part of the geographic majority or the linguistic and cultural minority. Sorting out the political and social arrangements constructed to bring a sense of order and security to our contact with others, needs to begin with an awareness of our own psychological and spiritual feelings. The more honest we are with ourselves, the more likely we are to be honest with others. Human beings don't come with guarantees so, for example, just because someone has a lot of formal education doesn't mean they can't be profoundly ignorant. Perhaps a worthy goal is to limit our ignorance to lack of exposure not lack of effort.

Robert D. Burgener BA, MA
Wivenhoe, England
Rockville, Maryland USA

Barbara J. Murphy, MD
Philadelphia, Pennsylvania
USA

Acknowledgments

First we need to acknowledge, perhaps not thank, whatever forces created the challenges that makes work like this necessary. Should the origin of our differences in language be traced back to a single event: an angry God punishing the hubris of people assuming they could build a tower to reach the heavens? Perhaps the origins of languages are as different as our perception of people who live on the other side of the mountain, across the sea or next door?

From an early morning walk in the Virginia countryside twenty years ago to an exchange of emails between my remote NATO basecamp in the Balkans in 1999 and his office at the Center for the Study of Mind and Human Interaction at the University of Virginia, Professor Vamik Volkan has been my friend and mentor. He has been my guide along a path of understanding conflict dynamics which he pioneered. Although for a much shorter period of time,(2005-2007) Professor Robert Hinshelwood of the Centre for Psychoanalytic Studies at Essex University in the UK provided insight, guidance and inspiration to dig deeper into the works of Melanie Klein and in the process, deeper into my own perceptions of myself and other people. His patience with an older post-graduate student, recently returned from military duty in Iraq filled with observations in search of a theory, will not be forgotten.

To our comrade- in- arms (and Barbara's son) retired Army Chief Warrant Officer Andrew Murphy for introducing us, and to all those who have inspired and frustrated us enough to take on the task of creating this window and mirror to see into the feelings of others, reflect on ourselves and, we hope, you our readers.

Chapter 1

How Language Becomes an Issue

Language is one of the more easily observed borders that divide otherwise similar human beings into groups of those who understand (what is being said) and those who don't. Graphic images created by professional advertising or local graffiti artists, the sounds emanating from our radios and televisions or the often mandatory options to select a language for navigating automated telephone answering services or conducting our credit card transactions, remind us borders with things we may not understand seem to have invaded our once predictable (understandable) neighborhoods.

Without dragging ourselves into the battle between creationism, represented by the Old Testament Biblical story of the Tower of Babel, versus evolution, with theories of spoken and written language development, the fact is there are over six thousand distinct languages used on planet Earth today. With the top twelve accounting for fifty per cent of the world's population, and a pattern of conflicts in the last decade that seem to always involve people who communicate in any one of the remaining five thousand nine-hundred and eighty-eight other languages, the different sets of sounds people in a community agree represent an object or thought, defines a formidable border when two or more communities meet. And the cost of crossing that border, with a guide, can be measured in terms of the revenue reported by associations of professional interpreters and translators which exceeded $8.6 billion dollars worldwide in 2005. Add another estimated $657 million the U.S. Department of Defense spends each year for language support to operations in Afghanistan, Iraq and other global hotspots with people who coincidently spoke some level of English and the target language, but had no training or professional credentials as interpreters or translators.

According to the U.S. Bureau of Labor Statistics, in 2007: "Employment of interpreters and translators is projected to increase 24 percent over the 2006-16 decade, much faster than the average for all occupations. This growth will be driven partly by strong demand in health care settings and work related to homeland security. Additionally, higher demand for interpreters and translators results directly from the broadening of international ties and the increase in the number of foreign language speakers in the United States. Both of these trends are expected to continue, contributing to relatively rapid growth in the number of jobs for interpreters and translators."

When European Union and United Nations Observers began arriving in the Balkans in 1991, they brought very few professional interpreters. The common language was English and their task was to recruit from the local populations at a time when many people feared for their own life and the lives of their families. Add to the emotional pressure, an evolving political-language dynamic in which Serbs and Croats now demanded interpreters to communicate their differences in unique phrases which were at once torn from what had been the common language of Yugoslavia, and at the same time represented deep, culturally rich and yet very different traditions. Croatian traditions, including language dating from the Tablet of Basca (1100 a.d.), were suppressed first in 1918, then to an even greater degree in Tito's post-WWII communist regime and finally by a man who may be remembered as the Butcher of the Balkans; the late Slobodan Milosevic.

It is not surprising then as the artificial political confederation of Southern Slavs (Yugo-Slavia), forged out of political convenience and sustained by fear and repression began to unravel, so did the contrived common language. The differences didn't begin in the 1990's, nor are the Serbs and Croats unique. As anyone who has served there, or followed that recent and dark chapter of human history knows, there are no contemporary saints in the Balkans. Cultural or ethnic amnesia seized all factions and conveniently ignored what had been shared in order to focus on what was different. This narcissism of minor differences, as Michael Ignatieff, quoting Sigmund Freud, calls it in *The Warrior's Honor, Ethnic War and the Modern Conscience*, (Ignatieff, 1997) forms a basis for dividing people. Language is merely one of the more obvious borders to be defended, or crossed at great risk to the interpreters, or those who simply want to learn another language. In an August 24th, 2008 New York Times article about the conflict between Russia and Georgia over South Ossetia, reporter Ellen Barry describes a young South Ossetian man who kept his language studies secret because of the animosity towards Georgians.

The 20th Century began with the first multi-national – Coalition – force of Westerners marching into a country half way round the world to rescue people targeted for ethnic cleansing. In this case, the group being targeted were not native born, wore different style clothing and spoke the same languages as the rescue force but not the local population. With the exception of some westernized locals, most of the native speaking population found themselves behind the same language and cultural barriers. Groups, most notably one called *I-ho-ch'un*, began to actively oppose not only the foreign territorial encroachment but the cultural and linguistic insults as well - a sign had appeared in a park near one of the foreign compounds warning "No Dogs or Chinamen" allowed. The English language translation of *I-ho-*

ch'un is "Righteous and Harmonious Fists" which for today's audiences would sound like an appropriate title for a martial arts film. In the early 1900's, the closest image was two bare-fisted fighters in a roped-off arena and thus an interpretation that gave us a name for the Boxer Rebellion in China.

The language requirements for that rescue force in 1900 were limited. The vocabulary necessary for communicating with the Chinese population was defined by the range and firepower of western weapons. In the year 2008, we find multi-national forces deployed around today's world, and we can hope their vocabulary has expanded to include non-lethal exchanges. Unfortunately the supply of ammunition usually exceeds the supply of qualified translators and interpreters.

On the political home front during the 2008 presidential campaign in the United States, the web-based organization known as *Media Matters for America* reported: "On the January 21 edition of the *CBS Evening News*, while discussing the upcoming Republican presidential primary in Florida, CBS correspondent Nancy Cordes reported that former New York City Mayor Rudy Giuliani "released a Spanish-language ad backing a catastrophic-insurance fund." Cordes went on to note that former Massachusetts Gov. Mitt Romney also "is airing his own ad in Spanish," and added that "Cuban-Americans make up a crucial 15 percent of Florida's GOP electorate and have not coalesced around a single candidate." However, at no point did Cordes note that Giuliani has said that English proficiency should be a requirement for citizenship, or that Romney has said that "English needs to be the language that is spoken in America" and that "[t]his is a nation which should never become a bilingual or multilingual nation."

On the same *Media* Matters page, Senator John McCain offered a political take on the issue of language stating, "I would like to remind you that we made treaties with Native Americans such as the Navajos in my state, where we respect their sovereignty and they use their native language in their deliberations. It's not a big deal. But Native Americans are important to me in my state.

"Everybody knows that English has to be learned if anyone ever wants to move up the economic ladder. That is obvious. (http://mediamatters.org/items/200801220011)

As we will discuss in later chapters, language is a big deal in the context of individual identity for those in the host and immigrant communities. The notion of a single economic ladder has also become less obvious. In Chapter 4 we provide examples of shadow economies. "In a 12 December 2006 feature story produced by BBC News entitled 'Lost in Translation' the reporter, Mark Easton, describes a section of east London known as Brick Lane where news about the UK appears only in the languages of the Bangladesh community - not English. In the opinion of a community activist this serves to further isolate ethnic Bangladeshi, many of whom have lived here for over twenty years, from the mainstream of their adopted country. (http://news.bbc.co.uk/go/em/fr/-/1/hi/uk/6174303.stm)

When couples from one culture and language adopt children from a different background, language can become an issue. Depending on the age of the child, a newborn compared to a 5 or 12 year old, the question of whether or not the adoptive parents should lean the child's language is a subject for debate. *Helium*, a forum for writers published the following responses to that question.

Carel Two-Eagle describes herself as "I'm an Indigenous Turtle Islander - an ITI. A name that is uniquely OURS, and one no can say was 'given' to us by members of some other culture. Vernacularly, that's an Indn, which is how we pronounce "Indian", partly to differentiate it from the name of those of the elephant culture. Not necessarily an "American" Indn, since this is only known as "America" because of a self-promoting Italian mapmaker, but definitely an aboriginal of this land. Sometimes I'll refer to myself as a horse-Indian; again, to differentiate it from elephant-Indian people" Raised in a Polish-English- Italian environment, she is in favor of new parents learning the child's language. "The fact that is more difficult for most adults to learn a new language than it is for children is not the over-riding concern in this matter. The over-riding concern should be the concern of the adoptive parents for the comfort - spiritual, psychological, emotional, not just physical - of the adopted child. Speaking to a child in the language of its birth culture goes a long way to promoting that comfort, and thus to reducing distress in the child, who suddenly finds itself torn from everything familiar and surrounded instead with strangeness on every aspect. If you find this disconcerting as an adult, think how it must feel to a child! No familiar facial features, no familiar sounds, no comfort from the familiar cadences and sounds of language... This cannot be kind!

In response to the same question, Jane Allyson is English, and she describes kindness in a different way. "Children need a lot of love and help to adjust to their new life. Concentrating on helping them to come to terms with their new language and a different life is far more important. Making an effort to learn anything that will help the child to feel happy and comforted is also essential to the bonding process, but you will instinctively know when the time is right to let go and concentrate on the present and say goodbye to the past."

Language triggers political and personal debates within ourselves and with strangers. There is no one answer that, once discovered, will fit each time the question of what to do about language comes up. It may be that hearing a foreign language reminds us of how our familiar, predictable world can change. The issue has more to do with how we handle change than the specifics of learning or not learning a language.

In this book we will discuss some of the theories concerning psychological processes such as the epistemophilic impulse, described by Melanie Klein (1928),and from *Identity:a Reader by* Paul du Gay, et.al. (2000) which offer insight into why we might find encounters with foreign languages uncomfortable. Based on first hand observations by an emergency medicine physician, Barbara Murphy,MD and a literature review of materials from international journals, we will then focus on examples from the health care field where mismanaged communication in a foreign language can become more than just uncomfortable. Flores, et.al. (2005), Bishoff, et.al.(2003) and others draw attention to poor choices of interpreters as a common denominator in cases of clinically significant errors involving patients and health care providers who don't share a common language.

Chapter 2

Language and Individual Identity

> *"Would you tell me, please, which way I ought to go from here?" [asked Alice]*
>
> *"That depends a good deal on where you want to get to," said the cat.*
>
> Alice in Wonderland,
> by Lewis Carroll (1865) chapter vi

When a person leaves their home and familiar surroundings they lose more than external objects, some say it seems like they have lost themselves. Finding the way back to yourself is a complicated, often painful journey. Drawing on the metaphor of Alice, refugees may well feel as if they have fallen - or been pushed - down an emotional rabbit hole and those involved with their therapeutic care, or any of us who come in contact with people from cultures other than our own, perhaps could take a clue from Carroll's character of the Cheshire cat by inquiring where it is we all want to get to.

In Robert Caper's 1988 work, Immaterial Facts: Freud's Discovery of Psychic Reality and Klein's Development of his

Work, he offers the following observation of Klein's (1946) paranoid-schizoid position. Caper writes:

She (Klein) used the term position rather than phase or stage because she wanted to describe a specific unconscious constellation of anxieties, defenses, and relationships to objects that, although originating at or shortly after birth, persist throughout all stages of life, as the mental life of the infant within the adult, unconsciously influencing our world views and conscious states of mind. (Caper 1988,176)

Perhaps in terms of external objects there is such a thing as the good language which we know and the bad language which in unknown? In 1928 Melanie Klein offered an epistemological context in which the infant's awareness of not knowing the answers to questions regarding relationships with parents and the infant's own sexuality become ingrained in the unconscious. "One of the most bitter grievances which we come upon in the unconscious is that these many overwhelming questions, which are apparently only partly conscious and even when conscious cannot yet be expressed in words, remain unanswered." (1975, 188). Klein draws our attention to the fact that this epistemophilic impulse has its origins in the first few years of life. "Thus his first questions go back beyond the beginnings of his understanding of speech." (ibid.)

According to Klein, "The early feeling of not knowing has manifold connections. It unites with the feeling of being incapable, impotent, which soon results from the Oedipus situation." (op.cit.) At this point in her career Klein still linked her observations closely to the framework established by Sigmund Freud's descriptions of life instincts, libido, which he expanded from exclusively sexual origins to include all psychic energy relating to hunger, thirst and other life sustaining activity (Freud, Beyond the Pleasure Principle 1920). Klein makes the connection between the child's feelings of impotence and fear of the consequences of not knowing to include, "numerous inhibitions of the epistemophilic impulse: for instance, the incapacity to learn foreign languages, and, further, hatred of those who speak a foreign tongue." (1975, 188)

Rose Edgcumbe (1981) draws on the work of L.S. Vigotsky to describe how the mother's response to an infant's first verbal attempts provides the feedback "facilitating the infant's shift from unintentional to intentional communication." (1981; 73) Edgcumbe describes how Vigotsky bridges the gap between the infant's physical world of pre-language communication and the propositional use of language.

Vigotsky (1966) answered this question by proposing that all human skills appear twice in their development: first they are constituted in interactions that take place between people (the intermental level); later they become the internal cognitive possession of one or both of these individuals (the intramental level). E.g., an infant cries in reaction to a state of discomfort, not with the intention of communicating to his mother. But she perceives him as if he were communicating; thus his ability to communicate exists through interaction with another. Feedback from others allows him to gain knowledge of his unintended ability, and it is transferred to the intramental level, becoming an ability over which he has control. (1981; 75)

Dependency on feedback to validate our communication leads some people to simply speak their own language in a louder, more forceful voice when encountering others who speak a foreign language in order to obtain an intermental reaction. On an intramental level, the absence of understandable response (feedback) or a response that must be filtered through someone else such as an interpreter, may represent a threat to control and a known sense of order and predictability required by the epistemophilic impulse.

In their sixth edition of An Introduction to Theories of Personality (2003), Hergenhahn and Olson refer to the work of John Dollard and Neal Miller and the importance of "images, perception and words cue-producing responses because they generally determine the next response in a sequence." (Hergenhahn and Olsen 2003, 329)

In keeping with Pavlov's behaviorist approach to language as a symbolic representation of reality expanding the repertoire of human response beyond the physical stimuli (sound, sensation or smell) that precede a biologically significant event (Pavlov's First Signal System) to include words such as love, peace and other abstract cues to real events (Pavlov's Second Signal System); Dollard and Miller described the function of cue-producing responses as the basis of reasoning and planning. As some of the examples involving the use of interpreters show, a great deal of reasoning and planning to the point of manipulation of reality are part of both the refugee attempt to access services and the care-givers delivery, or denial of those services. (see Temple, Chapter 6)

Within the theoretical realm of discourse analysis, consider the terms transactional and interactional. Transactional discourse exists for the purpose of conveying "factual or propositional information" and interactional discourse, is described as "language to establish and maintain social relationships." (Brown and Yule 1983:1-3). We will explore examples of how those functions are sometimes blurred in the process of moving between languages and how the parties involved influence each other.

Sigmund Freud (1937:263) offers his observations of "Constructions in Analysis" describing a patient's negative response to the analyst's construction because, in Freud's view it "points to incompleteness; there can be no doubt that the construction has not told him everything." Susan Budd and Richard Rusbridger (2005) take this further as they describe Freud's discussion of the complexity of recovered memory noting, "the telling of the tale lends its own logic; that the tale varies, acquires a narrative soothing, is a construction rather than a reconstruction." If the refugee 'recalls' a traumatic event - either from the original conflict that left them homeless or as the result of subsequent events whilst transiting the gaps between displaced person and asylum seeker - we must consider that all or part of the discourse may be a construction and not a directly recalled (unembellished) fact. If the interpreter helps, or is perceived to be helping in a shared construction of an event, the role of language as a manipulator takes center stage.

For the purpose of focusing our search for the role of language in the communication process between groups that are foreign to each other, we will consider the issue of identity. When a person leaves the country of their birth language they are likely to encounter examples of transactional language in the official forms and verbal interviews used by agencies in the receiving country. The pivotal importance of clear language exchanges at this stage cannot by over emphasized! When Kelly (1979) discusses how in moving from one language to another in literature there is a temptation to view "Translation as Creation," adding "Accuracy was clearly in the eye and emotions of the beholder." (Kelley,1979:45). If the 'beholder' is also a gatekeeper to asylum status, benefits and services, separating fact from creation is critical for everyone.

Once within the territory of their new country, social or interactional discourse then defines one aspect of how refugees or any group might be included or excluded by the host language community. As we will see in the following examples, language may also be used by any minority group to establish or preserve their identity and exclude others.

What is often a myth of assimilation and integration which fills much of émigré and refugee literature is examined in a different light by Elizabeth Loentz in her essay on pre-WWII Jews in Germany and by William Labov in his 1972 book, *Language in the Inner City*. In both cases, the authors are not describing groups that would qualify as displaced persons or refugees, yet they are isolated from the mainstream because of their unique language. Language intended to function as a social (interactional) as much as an information (transactional) medium. (Brown & Yule) A common denominator is that we are all subject to being judged by how we speak. "It is impossible for an Englishman to open his mouth without making some other

Englishman hate or despise him." (George Bernard Shaw Pygmalion, 1912).

How we use and are received or rejected by our spoken and written language provides a perspective that reminds us these social criteria do not exist as a unique aspect of being in a foreign environment, nor do they afford any special rights to the speaker in terms of what can be expected from the listener. I will explore this aspect later under the category of demands.

Loentz (2006), describes the use of language which has been Creolized, a reference to the hybrid language of the West Indies where "creoles, patois and Black English decanter, destabilize and carnavalize the linguistic domination of 'English.'" (Mercer, 1988:57 in Hinnekamp 2003). Hybrid language is referenced here to illustrate how, in the role of interactional discourse, it can be used as a means of group and individual identity to establish cohesiveness or challenge and rebel against the overwhelming forces of a host culture and language. Loentz (2006) provides the following example: "Minority and immigrant Germans' embrace of the derogatory term Kanake as a self-ascription and of the low-status ethnolect Kanak Sprak has been compared to US rappers' combative use of "niggah" and Black English." Loentz continues, "It demonstrates also how non-minority and non-Jewish Germans have used Yiddish and Kanak Sprak in literature, theatre, film, and popular culture to re-inscribe ethnic difference, especially at times when minorities and Jews were becoming indistinguishable from non-minority Germans (italics added)."(Loentz, 2006)

The term 'Kanak,' borrowed from the Hawaiian language kanaka maoli (kana: person) originally referred to Melanesians living in New Caledonia. It was no doubt introduced in Europe by French and German missionaries returning from the South Pacific, but soon took on a discriminatory role as a designation for any foreigner. (Encarta World English Dictionary, 1999; 980)

If Kanak Sprak or Yiddish were the sounds of being Jewish much as wearing a yarmulke provided a visible link, then for some individuals who grew up in secular neighbourhoods with none of those trappings and felt themselves part of a German culture, the news that the religion of their grandparents defined them as undesirables was devastating. As with many refugees today, they shared a common fate of having their identity imposed by a political process in which they could not participate. (Pegelow, 2006). When health care providers, or any group working in a foreign language environment, create expectations of dual language proficiency based on incomplete information about a person's ethnic heritage, they are imposing an identity on the would be interpreter that suits their needs and timetable, not reality. (Temple, 2006;5)

Labov (1972, 7:258-259) presents "The Linguistic Consequences of Being a Lame" in Black English Vernacular (BEV) where, "to be lame means to be outside the central group and it's culture." The opposite of 'lame' is to be 'hip' as in someone who 'hangs out' with street culture and is able to participate in the vernacular of that culture. For those of us who are outsiders to the Black or refugee experience, it is worth noting the term 'lame' is most often applied to individuals who might otherwise seem to be members of a group, but who are described by Labov as being "typical informants made available to investigators who study non-standard language in schools, recreation centers, and homes with the help of adults who control these institutions." (ibid: 7:255). When a person is recruited or offered up as a potential interpreter, the tendency to accept that person, often without the capacity to verify their vocabulary or interpreting skills, sets the stage for some often very bad outcomes. Will that person feel 'lame' if they must admit they don't know a specific word or phrase in both languages or will they choose to appear 'hip' to both their comrades and the outsiders by faking a false fluency? (Flores, et.al. 2003,8,10)

In choosing our perspective for viewing the refugee or host nation experience, we must be wary of choices based on convenience such as spoken or written language we understand versus language we must learn or experience through surrogate sources such as interpretation and translation.

Wearing clothing appropriate to the larger group or making an effort to lose an accent from your speech pattern are attempts within a racial grouping to reflect adaptation. Skin color or the shape of particular facial features, limit this option. The clothing worn by civilian interpreters working with the military is likely to be the uniform of that country serving both as notice to their ethnic brethren and a source of reassurance to the military personnel that the interpreters know which side they are on. Further exploration into the concept of negative identity which would suggest that a person feels safe in knowing who they are because they know who they are not, or extending that to the situation of the interpreter who wants to make the person paying their salary feel safe would require exploration of Donald Winnicott's "False Self," for example, which is beyond the scope of this paper.

I began this presentation with a theoretical framework for verbal behavior using Klein's observations of the epistemological motivation for knowing, rooted in instinctual drives for survival which find expression before the infant has a command of words. Our capacity to understand how words are used begins not with the symbols, but the personal perception of reality that motivates the choice of particular symbols. Brown and Yule provide a functional view of discourse analysis and finally I have presented several perspectives on how language serves to include and/or exclude in the process of establishing or defending our sense of identity.

Language creates a role for itself because it is an easily observed container of differences between those who understand and those who do not, exposing elements of compassion and prejudice, confidence and insecurity reflected in the actions of individuals and public policy. The outcome of contact between human beings is often determined by the capacity to communicate needs, desires and demands. Instinctual needs, embellished by a range of emotional forces become desires and together are finally expressed within a legal and social framework as demands. These theoretical elements provide a backdrop and undercurrent for the more observable aspects of both what is said and what is heard.

Chapter 3

Language as a Way Others See Us

Understanding how identity may be lost and alternatives for sustainable recovery found, constitute fundamental elements in the therapeutic dimension of helping refugees and the host nation communities where they are received. It is naïve to the point of being counterproductive if the subject of identity, in this context, is approached from only one perspective: excluding either the refugee or the host nation citizens as if they existed in isolation.

"To suggest that minorities are psychologically reactive to intercultural contact and that dominant groups are not almost implies that minority people are a different species of psychological being, one distinct from the majority. This is one step down the road to racism." (Johansen 2002 in Rudmin, 2003:5)

The issue of identity encompasses internal psychological dynamics of how we see ourselves along with socio-political dimensions of how we are seen, and our lives regulated, by others. As we examine the process of acculturation, keep in mind the role of the host community: "Whether immigrants can use the adaptation strategy they like best does not depend solely on themselves." (Oudenhoven 1997:1004).

Ephraim Nimni reminds us, "it is misleading to think of ethnic groups and nations as homogeneous communities. More often than not, the identity and prescriptive characteristics of the group is the outcome of internal power struggles in the interaction with the 'outside' world." (2007). It is this "multidimensional and social construction of refugeedom" (Papadopoulos, TV 900 Course Description 2006-2007) that challenges all of us and our notions of melting pots and what constitutes a nation of immigrants.

The relationship between identification, assimilation and integration thus suggests that if an individual identifies with his own group, he will hold favourable attitudes towards integration; on the other hand, if he identifies with the host society, he should favour assimilation. (Sommerlad, Berry eds. 1970).

Keeping or discarding certain articles of traditional clothing, for example, may be experienced by the host country citizens as a choice made by the refugee to demonstrate his or her willingness to integrate into the norms of dress reflected in the host country. That same gesture may be seen by fellow refugees as a rejection of them and the birth values they arrived with. Keeping the ethnic article of clothing can be interpreted by both communities as an act of defiance to display the refugee's separate identity. Sommerlad and Berry note that among "marginal groups" studied by Glaser (1958, 31-40) and Medding (1968) identification as "Both" (host society and ethnic group) was reported during the transitional stage, "since it enables them to retain identity with both groups." Rudmin (2003) reported "impossible results" revealed during a thesis study of Somali acculturation in Norway with respondents "endorsing two, three or four types of acculturation." (Merametdjian, 1995 in Rudmin 2003:41). We will discuss how the term acculturation incorporates the concepts of integration and assimilation later in this essay.

In a therapeutic context what can be learned from these examples is that the choices are far more complex than accepting or rejecting one option in favour of another. The necessity of making choices about things that had previously been accepted without discussion, as a natural part of the person's identity, brings issues of control over the decision making process to conscious levels that may not have been previously experienced. We will explore this concept further in the section dealing with the work of Wilford Bion.

For purposes of this essay I ask you to focus on the key word forced when describing individuals and groups who arrive in a foreign country under other than voluntary circumstances. The term forced is an adjective describing an action which is "unnatural; unspontaneous, done or provided under compulsion, carried out as an emergency." (Chambers, Encyclopedic English Dictionary, Oxford English Dictionary) and most revealing, "compelled: done because somebody who has power requires it" (Encarta ® World English Dictionary).

In these definitions we find two central themes. First, the personally destabilizing effect of an action not of the individuals' choosing, either in terms of timing or outcome and second, acknowledgement of an outside power that compels the action.

Refugees are not economic sojourners, as described by Oudenhoven (1997:996) in his study of Moroccans and Turks in the Netherlands or Brubaker "long term residents" in France or Germany (1998:134). While many of the emotions associated with loss of identity may be shared by anyone who finds themselves in circumstances where they may be defined as a foreigner, the process by which that designation is arrived at is significant. The migrant worker or "economic sojourner" may express feelings that they were "forced" by economic circumstances to leave their homeland, or drawn by the promise of a better life elsewhere, however their decision differs from that of the refugee because in the most basic sense, the refugee does not have a choice, or time to prepare.

The host nation response may be grounded in xenophobic perceptions of all foreigners or a misperception that lumps refugees and migrant workers into the same mix. The result in terms of distorting the identity of minorities and the process of how minorities regain or establish an identity remains the same. Hateful slogans urging people to "assimilate, integrate or go home" must have a very hurtful echo for those who have no homes. The specific political circumstances of civil war or ethnic cleansing vary along with what constitutes the well-founded fear that keeps them from returning: "loss of home is the only condition that all refugees share," (Papadopoulos, 2002:9)

When I submitted the original abstract for this essay along with the title: Assimilation - Integration: Complimentary or Contradictory Approaches with Refugees, I thought it would communicate an understanding of assimilation and integration as options that did not have to be mutually exclusive. What I have learned in the course of my research for this essay is that the terms "Assimilation" and "Integration" which I grew up with in the Midwest of America in the 1950's don't exist anymore - and perhaps never did, or more to the point, never should have. Rogers Brubaker's comment about the iconoclastic preface written by Nathan Glazer and Daniel Patrick Moynihan for their 1963 classic *Beyond the Melting Pot* in which they said there was, in fact, a failure to melt, struck me as only partially accurate. (Brubaker 2003:531) In 1963, I left a small farming town to attend a state university in an equally small, but far more cosmopolitan academic community where I met more Asians, Middle-Easterners, African-Americans and people of different religions than ever before in my life until that point. My ticket to sit at what I thought was a far more fascinating "foreign student" table at the student union was not my father's Swiss family name, which had helped me get into one of the Greek-lettered social fraternities, but the fact that my mother was French and Cherokee Indian - Native American as we now call it in more politically correct terms. The problem was my grandparents had died many years before I was born, and the subject of being Indian was taboo as far as my mother was concerned. Being Indian in the still segregated mid-west in those days was the same as being Negro before the Civil Rights movement. I had been blended and melted into the great WASP (White, Anglo-Saxon, Protestant) middle class and like a lot of third or fourth generation refugees from other cultures, the resources for reconstructing that part of my identity were lost.

We are stuck with the title, but I hope the contents, as they have evolved during my research, will provide the broader context this subject deserves.

Using the Fourfold Paradigm of acculturation described by Rudmin (2003) with credit to John Berry and his associates (1970, 1972, 1974, 1977) what happened to me and my mother's family would be diagrammed as (-M+D). When a minority culture (M) comes into repeated contact with a dominant culture (D) the outcome can be described as assimilation (-M+D) where the minority language and culture are completely absorbed into the dominant culture and come to possess (and are possessed by) a sameness. If the minority maintains its differences, the equation for separation is written (+M-D). When two cultures meet and there is not a quantifiable or enforceable measure of dominance, the two may coexist (biculturalism) in a state of integration expressed as (+M+D). The fourth and final option is described as marginalization (-M-D) which as the (-) negative sign suggests, means neither side profits or loses from the contact.

Rudmin's detailed examination and critical analysis of Berry's Fourfold model as acculturation taxonomy includes analysis of its failure to accommodate categories of individual behaviour or hold up under various forms of statistical analysis and points to the fact that there have been at least seven earlier versions of fourfold acculturation typologies. (2003:8). In terms of establishing how the frames of reference for minority and dominant are determined, I found Rudmin's historical notes of particular interest:

According to the Oxford English Dictionary, the word acculturation was first used in English text in 1880 by J.W. Powell to describe changes in Native American languages: 'The force of acculturation under the overwhelming presence of millions [of Europeans] has wrought great changes' (p.46)" (Rudmin 2003:10).

Later Rudmin goes on to describe the work of G. Stanley Hall (1904), "the first avowed psychologist to write about acculturation" who observed that "soon after their 1620 landing in Plymouth, settlers were trying to acculturate the Native Peoples, even though the settlers were the immigrant minority." (italics added.) (Rudmin 2003:10,11).

The presence of the European settlers may have been overwhelming in their minds and reflected in their writings, but it was their cultural arrogance, not a numeric superiority that defined the outcome for North America and its indigenous peoples. What might the world look like today if the "differentialist" turn described by Brubaker had taken place in the 1680's or 1880's instead of the 1980's? In their examinations of pre-Colonial America, Authors Samuel Payne, Jr. (1996) and Bruce Johansen (1990) describe various degrees of cooperation between native American Indian tribes, such as the Iroquois League which traced its roots to the 12th century, with each other and then the new arrivals.

Was it a matter within the personality mix, to borrow a phrase from Thomas and Znaniecki whose work will be discussed later, of some North American Bohemians and Philistines, wearing feathers whilst other wore wigs? We may never know the full story because the "wigs" wrote the history. When those concerned with preserving differences (separation, heterogeneity) are positioned as an opposing force to those advocating assimilation (complete absorption, homogeneity) the winner defines the political, social and individual world as we know it. If we accept those polarities as the only options for defining ourselves, then there may not be any winners.

At the turn of the last century, Brubaker observes in the pre-WWI polices of German Chancellor Bismarck, "The (Prussian) state feared that Polish immigrants would strengthen the ethnically Polish at the expense of the ethnically German element in the Prussian East." (Brubaker, 1998). Are those sentiments echoed in a British newspaper, the Daily Mail, editorial by Sir Andrew Green (April 21, 2007:14) under the headline "The Great Deception"? With regard to Britain being a nation of immigrants, Sir Andrew says, "Tosh! Immigration today is 25 times higher than at any time in history." I wonder if words similar to those were spoken around the council fires of the Iroquois League or amongst other native American tribal leaders with regard to the influx of Europeans and their African slaves? The political dimensions of identity, expressed as citizenship, whether as a subject of sensational tabloid headlines or described in legal or academic terms by authors such as Yasemin Soysal in *Postnational Citizenship: Reconfiguring the Familiar Terrain*, Brubaker, Nimni and others, has its own complexity which I will not explore any further in this particular essay.

The dynamics of how political and social decisions are made are best appreciated from the perspective of individuals through the study of human behaviour (external) and our psychological makeup (internal). Wilford Bion describes a union of people brought, and held, together by some shared assumption:

As far as the group is concerned, the basic assumption is essentially a tacit assumption. Individuals behave as if they were aware of assumption, and it is for this reason that the interpretation of the basic assumption carries conviction. (1961:93,94)

If we consider the alignment options available within the Fourfold process of acculturation (-M+D, +M-D, +M+D and -M-D) from the perspective of Bion's basic assumptions of Fight-Flight (baF), Pairing (baP) or Dependency (baD), the role of the individual in articulating public policies and personal choices can be appreciated as the controlling element. As Bion described it:

We thus have a situation in which the individuals behave as if they were conscious, as individuals of the basic assumption, but unconscious of it as members of the group. This is as it should be: the group has not a conscious; and it is not articulate; it is left to the individual to be both.(1961:94)

The official history of policy and laws regarding how cultures regulate contact with each other goes back 1700 years before Christ to Babylon, followed by the Persians, Greeks and Romans; but one of the first psychological descriptions to emerge came from a 1918 study by William Thomas, and Florian Znaniecki, titled The Polish peasant in Europe and America. If we look back at some of the examples cited in this essay, or over our shoulder at the people we work with everyday, the personalities they describe will be quite recognizable.

Three types of acculturation arise from personality differences in the two biological dimensions of fear and curiosity. First, individuals with Bohemian personalities are low in fear and high in curiosity, causing them to seek change and to adapt to any social context but at a cost of personality disorganization. Second, individuals with Philistine personalities are high in fear and low in curiosity. Thus, holding fast to social traditions and rejecting modernity, the Philistine-type personality becomes non adaptive in new sociocultural contexts.

Finally, individuals with creative personalities have a balance of fear and curiosity and thus seek controlled and coherent change. By maintaining but modifying the schemas of the minority culture to adapt to the dominating pressures of modernity, the creative-type personality can “widen the control of his environment” and “adapt to his purposes a continually increasing sphere of social reality.” (Thomas, Znaniecki 1918, Vol.2 p.1856)

In the interest of maintaining an accurate perspective, perhaps an occasional glance in the mirror would be useful?

If we accept these Bohemian, Philistine and Creative personality types as one means to describe how individuals adapt to or resist change, then it might be useful to consider how society organizes our options for change. In 1964, University of Massachusetts sociologist Milton Gordon wrote Assimilation in American Life The Role of Race, Religion and National Origins, which outlined the stages or phases of the process by which one group was absorbed by another. Six years later, in 1970, Elizabeth Sommerlad and John Berry (of Fourfold fame) used Gordon’s analysis of assimilation as a basis for describing how integration differs from assimilation. Quoting an Australian study (London, 1967:340 in Sommerlad, Berry 1970:24) “Integration implies interaction between the minority group and host society with a resulting change in the cultural amalgam, but without individual loss of cultural identity.” The “subjective factor of identification” was considered significant, to the point of being what the authors called, “the key difference between assimilation and integration.”(1970:24)

The issue of identity in marginalization (-M-D), discussed in *Acculturation Attitudes in Plural Societies* (Berry, Kim, Power, Young and Bujaki, 1989:188), describes “acculturative stress” (Berry, Annis, 1974; Berry, Kim, Minde and Mok, 1987) as “feelings of alienation, loss of identity,” which emerge when “groups lose cultural and psychological contact with both their traditional culture and the larger society.” (1989:188). Evidence of alienation from the larger society may be found in the examples the individuals who planned and carried out the terrorist bombings in London on July 7, 2005 (7/7) which will be discussed later in this essay.

Gordon's Assimilation Variables

Type or Stage of Subprocess or Condition	*Assimilation*	*Special Term*
Change of cultural patterns to those of host society	Cultural or behavioural	Acculturation
Large-scale entrance into cliques, Clubs, and institutions of host Society, on primary group level	Structural	None
Large-scale intermarriage	Maritial	Amalgamation
Development of sense of people-hood based exclusively on host society assimilation	Identificational	None
Absence of prejudice attitude Absence of discrimination behavior	Receptional	None
Absence of value and power conflict	Civic	None
(Gordon, 1964:71)		

The differences between assimilation = total and complete absorption to the point that characteristics of the minority culture disappear in the host culture, and integration = interaction with accommodation for differences are described as follows:

"Firstly, and most important of all, the identifying stage, whereby the minority group changes its identity to that of the host society does not occur." Finally, "assimilation is nearly always a unilateral process whereas integration stresses the mutual contributions and adjustments both groups have to make." (ibid.)

The question then is not whether assimilation and integration are complimentary or contradictory approaches, the basic question is how those terms are perceived and understood in the context of the minority and dominant cultures and by the individual's making the choices. A more appropriate question is, at what price do we buy into assimilation or integration? The second component in terms of therapeutic care for refugees involves acknowledgement that these are not the only approaches.

While timelines, models and stages may suggest a linear progression from one event to another, in fact the behavioural elements of identification, regression and projection, make any collection of experiences a fluid dynamic in which we may move, often in a disoriented way, from one level of consciousness and assumptions to another. Whichever analytical tool we employ, Berry's, Gordon's or Papadopoulos', if in the case of refugees from ethnic conflict, the pre-event stage contained examples of assimilation and integration, such as marriage between people of different religious and ethnic backgrounds or the sharing of social and work functions; then the betrayal of the trust associated with assimilation and integration which took take place during the initial period, taints them as viable options in later phases of "Survival" and "Adjustment."

Separation (+M-D) and marginalization (-M-D) often coexist with assimilation and integration within the same groups because the Bohemian, Philistine and Creative personality types described by Thomas and Znaniecki (1918) are alive and well and causing trouble.

Rudmin reminds us, "History is one way to establish enough distance from the social dynamics and academic fashions of a contemporary paradigm to observe its trajectory from the past, through the present, and into the future." (2003:41). As the official commissions, contemporary news commentators and pundits exhaust ways to get a few more stories out of the blame game surrounding 9/11 or 7/7, a U.S. presidential commission has issued The Iraq Study Group Report (2006) which offers an important insight: "Good policy is difficult to make when information is systematically collected in a way that minimizes its discrepancy with policy goals." (Baker, Hamilton, et al 2006:95)

At the individual level this would be called prejudice, for large groups like governments, it is part of the priority-setting assumptions used to create the rules and regulations we live by. But how do we cope with bad things that happen to good people who try to live by those community rules? It is necessary to look at the collateral damage caused by events such as Pearl Harbour, 9/11 or 7/7, measured by the lives of people in minority communities who were somewhere along their personal assimilation - integration journey when these events took place. The American government interred Japanese-American citizens in camps during World War II. How assimilated or integrated was that?

The British press continues to reveal bits and pieces of the background of the 7/7 plotters with photo captions that read: "RINGLEADER WHO COULD HAVE BOWLED FOR ENGLAND" and "MODEL WHO SOLD ALCOHOL IN TESCO." (Rayner, Williams in Daily Mail, 1 May 2007:2). The disturbing aspect of their stories is the common thread binding these young men who were educated and spent the bulk of their adult lives in Britain, yet seem to have found their identity in the radical policies of foreign organizations. Such accounts are part of the acculturation environment today as they probably were in ancient times or in the European discovery of the Americas and they demand consideration in any therapeutic approach.

The history books of the future may well look back and note how this chapter of human activity was written by those who remained separate and marginalized from the social and political communities where they lived.

Conclusion: While assimilation and integration may seem to represent broader long term goals, in fact they each contain the stuff of which day-to-day life is experienced by both refugees and the host community, not as abstract goals, but often harsh reality.

The psychosocial and political elements that define belonging and status within a group, serve to define assimilation and integration for those outside the group, often confronting them as they attempt to move from an established identity associated with their country of origin towards some alternate means of recognizing themselves in a new community. Do you enter the new group as your old self and demand recognition or do you try to adopt the language and dress codes, for example, in order to blend in and hope you are not recognized? "Passing is comparable with assimilation." (Taylor, Moghdamm 1994 in Oudenhoven et al 1998:998)

In the final analysis there will always be forces that seem beyond our control. We cannot depend on the accuracy of the weather or political forecasts and so with wisdom passed down in folklore and children's stories, we learn that what cannot be predicted, must be accommodated. The strength with which we hold onto our sense of identity may well be the key factor in determining our capacity for such accommodation.

Chapter 4

Building Bridges or Walls with Words

"... we have to defend ourselves with a powerful weapon. This weapon from time immemorial has been art, especially narrative art: literature, poesy, film. ..."

Edgar Reitz (1998) German film maker.

Language, in its many forms, will be explored from a sociological and linguistic (discourse analysis) perspective. My goal is to create a thread of many colours connecting different pieces of intellectual, and ordinary, fabric into a functional patch-quilt of arguments and evidence. It will not be a delicate tapestry of details but rather a simple, functional collection of approaches that may serve as tools for anyone wishing to build on opportunities, tear down obstacles or engage in a critical discourse concerning the refugee experience and experiencing refugees. In the dynamic process of attaching arguments and evidence the thread blends with some pieces and contrasts with others in a succession of complimentary and clashing juxtapositions which we have come to recognize as the multiple contexts of a politically created and defined dimension of human relations: being a refugee or a citizen in a country where refugees flee.

Human beings and the social order or disorder they create present multiple aspects for study and this essay will attempt to expand on examples of literature and film by introducing elements from discourse analysis, sociology and psychoanalysis. Each of these elements is worthy of exploration as a single course of investigation into caring for refugees, however in keeping with my understanding of the requirement for "demonstrating the multidisciplinary nature of Refugee Care" (Papadopoulos, R. Masters in Refugee Care M35 Essay Assignments, 2006-2007) I will discuss sociology and linguistics, with a focus on discourse analysis.

The central fibre of our connecting thread is communication through language and images. Blending ideas of memory as construction or reconstruction (Freud, 1937:), transactional and interactional discourse (Brown, Yule 1983), and interpretation as transmission or creation (Kelly 1979), our thread expands its capacity to hold together what we may seem to be contradictory examples of language as bridge and barrier between the refugee and host nation communities. (Easton,. BBC Feature, 2006). My arguments and evidence encompass both the care-giver and care-receiver along with the larger socio-political container of public opinion.

Many of our most reliable opinions are formed by direct contact with subjects using all our senses, intellect and instincts. The validity of that approach has not, however, prevented mankind from using images such as photographs, film and video or spoken and written language to create a body of surrogate experiences of events and people with whom we have no direct contact. The ability to draw on the positive aspects of such surrogate resources whilst recognizing the inherent dangers of intentional or circumstantial manipulation,needs to be a part of a caregivers' approach.

University of Cambridge social anthropologist Esther Goody discusses the capacity of language to move us beyond the limits of sensory input: "Homo sapiens depends on learning rather than instinct for the transmission of adaptations between generations; effective learning on the scale necessary for the transmission of culturally developed adaptations depends on language." (Goody, ed. 1995:9) Beyond simply satisfying a brief curiosity, second-hand observations by others are incorporated at the conscious and unconscious level with our own direct experience forming our opinions and shaping our actions. "We act according to the way we conceive things," (Lakoff and Johnson, 1980:5) and it is through this conceptual process of describing ourselves and others that we will discuss the context of the refugee experience and the impact on care-givers (experiencing refugees). Language, in whatever form we experience it, offers both a window through which we may peer into the lives of others and a mirror revealing an image of ourselves.

This essay will draw on examples of transactional discourse for the purpose of conveying "factual or propositional information" and interactional discourse, described by the authors as "language to establish and maintain social relationships." (Brown and Yule 1983:1-3). We will consider examples found in the electronic media and traditional literature, to illustrate the way refugees are portrayed by the host community and by the refugees themselves. In shaping the "dominant- discourse-vacuum" within the refugee community which can be highly polarized along political lines (Kuscu and Papadopoulos, in Papadopoulos, ed. 2002:260, 61) it is useful to consider at what point the conversation turns from factual information (transactional) and becomes a device for social bonding (interactional) around a shared trauma, often embellished with an ethnic, religious or political agenda.

The same may be true for oral histories or other literary forms - are they documentaries or docudramas? Are the stories constructed by mainstream publishers or broadcasters reflecting facts or are the 'facts' arranged to reflect a deeper theme of social angst about survival on the economic and culture level? Is anyone telling us everything or is it up to us, in our interdisciplinary approach, to piece together the different perspectives into something we can hold as our version of truth?

In this essay we will explore a broader definition of literature beyond what most dictionaries list as a first category of "written material of high quality, valued for its language and content such as novels, poem and plays." (Chambers Encyclopedic English Dictionary, 1994.74) or subsequent categories which cast a broader net to include "the whole body of written works of a particular country or period." (ibid. 74).

Taking Marshall McLuhan's book The Gutenberg Galaxy: The Making of Typographic Man (1962) as a starting point for the ways and means reproduction and dissemination practices evolve, changing how audience's access information, this essay will include television and internet sources as part of the medium of surrogate experience which allow us to engage in remote witnessing of events. Witnessing through the camera lens or words of others, can mean becoming what Michael Ignatieff describes in The Warrior's Honor, as "voyeurs of the suffering of others, tourist amid their landscapes of anguish." (Ignatieff 1997:11) Providing care for people who are refugees requires making connections where ever we can find them. Understanding how language is experienced can be one of those connections.

Once within the territory of their new country, social or interactional discourse then defines one aspect of how refugees or any group might be included or excluded by the host language community. As we will see in the following examples, language may also be used by any minority group to establish or preserve their identity and exclude others.

Does the language of literature in any form define the context of how we experience others? The obvious answer is yes, if we can't understand the language thus blocking both the information content and social interactional dimension. Can learning the language of another group actually make you a member of that group and expand your circle of experiences? The answer again may seem obvious, but in fact it is far more complicated than just matching words and meanings. For some, the ability to read and understand a foreign language allows them to participate in witnessing the refugee experience through the surrogate vehicles of literature or foreign news broadcasts without leaving their secure, native community comfort zone. The refugee or other foreigner living in a discomfort zone defined by words they don't completely understand may forge ahead in overcoming yet another obstacle (language) or rely on others to tell them what 'they' said. In a reversal of the first scenario described above, some refugee communities insulate themselves with news about the host country provided in their birth language.

The issue of which language a writer from the emigrant or refugee community should use in telling the story of that community is determined by the practical necessities of finding a publisher in the host country and a philosophical motivation, described by Clare Finburgh: "if literature's written in the language of the oppressor, the oppressor can understand and recognize the protest." In another example of a writer almost unknown to the countrymen of his birth language, Nigerian Nobel Laureate Wole Soyinka wrote in English, rather than his native Yoruba. That choice made his work available throughout the world in part because of the ease of translation from English to other languages compared to more limited access to Yoruba. (Finburgh, C. 2006, 2007 Seminar Notes on Driss Chraibi)

It can be argued that literature in translation is as much a barrier as it is a bridge. In a 12 December 2006 feature story produced by BBC News entitled 'Lost in Translation' the reporter, Mark Easton, describes a section of east London known as Brick Lane where news about the UK appears only in the languages of the Bangladesh community - not English. In the opinion of a community activist this serves to further isolate ethnic Bangladeshi, many of whom have lived here for over twenty years, from the mainstream of their adopted country. (http://news.bbc.co.uk/go/em/fr/-/1/hi/uk/6174303.stm)

London and the UK are not an isolated example. In a 2003 working paper for the Forced Migration and Refugee Studies program at The American University in Cairo, Daniele Calvani describes a similar situation among Sudanese refugees in Egypt. The refugees, now in their forties and fifties have been living in Cairo for six or seven years yet, according to Calvani, "They all adamantly declared that their Sudanese Arabic is unchanged." (Calvani 2003:9) and added they "reject any unnecessary contact with Egyptians." (ibid: 10).

Taking up the issue of belonging, Hinnenkamp (2003) and Cheesman (2002) present the use of a hybrid language by young people in Germany as a means of creating an identity that embraces their differences from the mainstream and at the same time excludes that mainstream in what became known as 'Kanak.-Chic.'

"In the later 1990s the writer Feridun Zaimoğlu seized the label `Kanak' to create himself as a spokesman for Turkish and other visibly non-German Germans who have been raised in the Federal Republic but have outsider status." (Cheesman 2002) According to Cheesman, Zaimoğlu was motivated, at least in part by "the patronising exoticism practised by promoters of `Migrantenliteratur', and the conformism of Turkish-German politicians and writers of `Gastarbeiterliteratur'." (ibid, 2002). We will visit the topic of migrant or 'guest worker' literature later in our discussion of Driss Chraibi's 1955 novel The Butts (Les Boucs). Boundaries established by language are perhaps most visible when we see the other language in large letters such as on a sign. If it is part of a store front window in a minority neighbourhood, that is to be expected and may be seen as part of the atmosphere defining 'their' space and comfort zone. But when the foreign language appears in our own neighbourhood, that's a different story.

The newspaper headline read: "A sign of things to come? English road gets directions in Polish" The story that followed described how, "A British local authority has erected road signs in Polish to cater for growing numbers of drivers from the eastern

European country." (Agence France-Presse, AFP, February 15, 2007) A discussion of road safety issues and the impact of European Union relaxed travel regulations led to the following comment, "The decision prompted an outcry from anti-politically correct campaigners who questioned the financial implications of erecting the bright yellow signs." (ibid). In the words of a conservative member of the British Parliament, "It's absolutely bonkers but what worries me is that once one council starts, others follow." (ibid). In Immanuel Kant's essay Perpetual Peace (1795) he sets forth a "Third Definitive Article for a Perpetual Peace" in which "The Law of World Citizenship Shall Be Limited to Conditions of Universal Hospitality." Kant goes on to describe those conditions in detail including the "right of a stranger not to be treated as an enemy when he arrives in the land of another," and notes the extension of hospitality and the withholding of hostility "is not the right to be a permanent visitor that one may demand." It is, Kant argues, "only a right of temporary sojourn." Are the long-term local inhabitants reacting to the foreign language signs as encroachment on their hospitality that suggest the visitors have overstayed their welcome and may be planning to remain even longer? Do we let the tourist get lost and eventually find their way to another community where they spend their holiday money just because we want to punish the emigrant or refugee who hasn't learned our language? The financial implications of language issues for both the refugee and host communities is not one of our selections for inclusion in this particular patch-quilt, but it is one of the realities. If we agree with Nietzsche that life is now, in this world and not in some hereafter, then the plight of refugees as wanderers, not belonging in their new location and unable or unwilling to return homeland, becomes easier to explain in terms of the violence and anger directed inward towards themselves and outwards at the points where they contact the 'other' communities. Defining home becomes a quest and a burden portrayed by chalk lines on the floor of an open barracks or a list of street addresses and apartment numbers as they are shuffled from place to place by various agencies, or the kindness of strangers.

Whether written or spoken, words used by refugees to describe themselves provide connections for others into the inner and outer worlds of those who may be seen as 'others'. Words used by members of the host nation to describe refugees offer insight not only into how the refugees are seen, but how the observers see themselves. When, for example, Driss Chraibi chose the title The Butts (Les Boucs) for his novel about North African migrants in France, he opened a floodgate of metaphors revealing how migrants see themselves as animals and how the French behave, not as shepherds, but as cattle drivers. In the use of derogatory language, calling outsiders names that reinforce images of being worthless, less than human, one community seeks to distance itself from the others. (Finburgh, C. 2006, 2007 Seminar Notes on Driss Chraibi)

In Chalabi's character of Isabelle, a French woman who befriends the migrants we are shown a resilience in coping with her own physical disability (we are told she has a bone condition as a result of being malnourished during the war) which is at once a levelling of the playing field, a credential for genuine compassion, and an indictment of the North African migrants lack of resilience in dealing with their social and emotional disabilities. When individuals and communities try to idealize or demonize others, in person or through surrogate devices such as literature and slogans, this behaviour might be investigated as an example of what Melanie Klein (1946) described as projective identification? We will come back to this point in a moment. First, perhaps if we consider our earlier examples from Goody on how "Homo sapiens depends on learning rather than instinct for the transmission of adaptations between generations; effective learning on the scale necessary for the transmission of culturally developed adaptations depends on language." (Goody, ed. 1995:11) along with Susan Budd and Richard Rusbridger (2005) description of Freud's discussion of the complexity of recovered memory where they noted , "the telling of the tale lends its own logic; that the tale varies, acquires a narrative soothing, is a construction rather than a reconstruction." We may begin to see a pattern in the refugee experience and in experiencing refugees that will help us understand how prejudice is transmitted from generation to generation. In caring for refugees some of us are insiders to the refugee community because of our own set of experiences and roots while others are outsiders in the sense that we come from one of the host nations. We cannot avoid being part of the transactions, but awareness of the complexity and multidisciplinary aspects of the information and emotional exchange can increase our understanding of the process.

If we return for a moment to the characters provided by Chraibi (1955) in The Butts (Les Boucs) there is an antagonistic quality in the relationship between his French woman Isabelle and the male North African migrant workers. In an exchange with one of the main Butt's characters, Waldik, Isabelle delivers a speech that contains sentiments reflecting the feelings of many who come in contact with refugees.

"Yes, I am ashamed to be European, but it is you North Africans yourselves that I condemn the most. You have always let yourselves be exploited, and you have always been exploited." (Chraibi 1955:116). Is it possible that people in the host nation project their guilt regarding the plight of refugees back onto the refugees for doing something to put themselves in that position? In exchanges between communities of real people, not fictional characters, is there an element from Melanie Klein's description of projective identification in which people split off their unwanted feelings and deposit them in others, waiting to see how the others will process those feelings? (Klein, 1946)

This piece of academic fabric belongs in our patch quilt, and may in fact be one of the fibres in our connecting thread because it contains all the colours of the motivation of writers seeking to put readers in the place of others, the use of language like Kanak-Sprak to hold onto identity. The emotions of empathy and alienation seem bound up in statements that claim we are able to accept some foreigners as individuals but reject them as a group. Klein tells us there is an "intrinsic need for integration" to bring the good objects back into ourselves.

Language and the process of surrogate experience may in fact be an example of how we relate to people who at first seem different than ourselves. The context of experiencing refugees and the refugee experience embodies the two essential components of this social dynamic. Each individual and community exists within a set of language and cultural boundaries, crossing those boundaries can bring out the best and worst in everyone.

We may look at others through a window and ourselves in a mirror. In each instance language as a window or mirror serves as a benign device through which we project our conscious and subconscious interpretation of what is seen and heard.

It is in the process of internalizing the various forms of communication and incorporating them into our opinions and actions those elements move from benign information into potent actions. Words may represent simple, observable facts as a matter of record (transactional discourse) with no further action required; or they can evoke a cascade of emotion, prompting us to do something (interactional discourse). Like the carpenter or glass maker, writers, and all of us, use words to craft ornate or simple devices intended to create an opportunity to look into the lives of others, or reveal reflections of ourselves.

In this essay we have followed an interdisciplinary thread of many colours connecting different pieces of academic fabric, provided by sources such as social anthropologist Ester Goody, linguists George Lakoff and Mark Johnson along with brief insights into the psychological processes of constructing memories revealed in the works of Sigmund Freud, as described by Susan Budd and Richard Rusbridger and projective identification by Melanie Klein. It has not been a delicate tapestry of details - I have sought to encourage the reader with more questions than answers - but rather a simple, functional collection of approaches that may serve as tools for anyone wishing to build on opportunities, tear down obstacles or engage in a critical discourse concerning the multidisciplinary nature of experiencing refugees and the refugee experience.

In my attempts at attaching arguments and evidence the thread blended and contrasted in a succession of complimentary (language as an interactional social medium) and clashing juxtapositions (language as a means of exclusion) which we have come to recognize as the multiple contexts of a politically created and defined dimension of human relations: being a refugee or a citizen in a country where refugees flee.

Acknowledgments

Clare Finburgh and Alan Cardew at the University of Essex for the original inspiration and subsequent counsel as I tried to patch together examples that would do some justice to the subject and their passion. Any success is due to their guidance, the failures are my own.

Chapter 5

Crossing the Language Boundaries in Medicine

This chapter presents some of the results of a literature search focused on the impact of communication errors between refugees and the health organizations responsible for providing care. Of the many aspects of human emotion ranging from personal identity to national pride and economic constraints involved in such international contact, we offer examples of how absence of a common language and poor choice of interpreters emerge as observable factors in failure to access services and incomplete or inaccurate delivery of care.

Concentrating on documented situations involving persons in need of medical services attempting to communicate those needs and health care providers processing that communication with questionable filters (language support) we see how the best possible outcome can remain elusive. The normally dyadic relationship between the person in need and the person capable of taking action to satisfy that need takes on a new and sometimes more complicated dimension with the introduction of a third person – the foreign language interpreter. Going back to our theoretical sources, the Oedipal conflict involving the child desiring a relationship with one parent but recognizing the other parent as an obstacle to attaining the desired intimacy creates a situation in which "what was important was not the impact of external reality or of internal impulse alone, but their sum." (Caper 1988, 39). I will pursue the dynamics of triadic relationships in a future paper but for now, our literature search provides examples such as this from Greenhalgh, T., Robb N., and Scambler G. Of University College London, UK:

Our data suggest that the preconditions for communicative action are rarely met in the interpreted consultation. The interpreter's presence makes a dyadic interaction into a triad, adding considerable complexity to the social situation and generating operational and technical challenges. Lack of trust, intense pressure of time, mismatch of agendas (biomedical versus lifeworld), firm expectations of a specific outcome (e.g. referral, prescription) and profound power imbalances all promote strategic action (i.e. speech that seeks consciously or unconsciously to manipulate an outcome) rather than communicative action (i.e. sincere efforts to achieve understanding, and reach consensus) by all parties. In consultations interpreted by family members (an option traditionally seen as 'second best' by policy makers), the social situation is very different. Family members are generally trusted, share the lifeworld agenda, and shift the power balance in the patient's favour. The interpreter occupies multiple social roles, including translator, interpersonal mediator, system mediator, educator, advocate, and link worker. The essence of professionalism in interpreting is shifting judiciously between these potentially conflicting roles. (Greenhalgh, Robb and Scrambler in Social Science Medicine. 2006 Sep; 63(5):1170-87.

The examples presented were obtained from multiple database searches using the key words: *medical interpretation error* and *communication barrier AND translation AND limited English proficiency* they contain many of the elements of personal vulnerabilities and public policy in pragmatic settings from emergency rooms and children's clinics to refugee screening centers .

In the case of refugees and asylum seekers we are dealing with a group of human beings who have been abruptly displaced from their homes and birth-language with no time to learn to speak a new language. The epistemological need to find some order and predictability in their situation (Klein, Dollard and Miller) is confounded by an inability to understand what is being said. Contact with foreign languages may begin almost immediately if international peacekeeping forces are inserted into a conflict and internally displaced segments of the population must turn to them for protection and survival.

Upon arrival in another country the legal term refugee and asylum seeker are applied to one aspect of the dynamics that will define the next period of their lives. In a Brookings Institution – University of Bern project describing the plight of the estimated 1.2 to 1.5 million Iraqi Refugees in the Syrian Arab Republic, Ashraf al-Khalidi, Sophia Hoffmann and Victor Tanner note:

Iraqis sought refuge in Syria, rather than in other countries, for a number of reasons: geographic proximity, simple entry requirements, easy access to services, common language, the low cost of living and often the presence of family or friends in Syria. (al-Khalidi, Hoffmann and Tanner 2007,1)

Access to an airline ticket has redefined 'geographic proximity' creating cultural and language contacts that are, at best, awkward for all involved. The political and social dynamics of competition for resources ranging from jobs to public services produce a volatile mixture of people fleeing from physical danger (refugees) and economic migrants seeking greater prosperity. For the people of the host country these foreigners are often lumped together, identified only by the difference in language. As the authors of the Brookings -Bern paper (2007) conclude, sharing a common language has not eliminated the other problems faced by any refugee community. Lack of a common language can create situations that are not only awkward – they can be deadly. When clinically significant errors in interpretation between languages lead to incomplete or misdiagnosis, then access to appropriate services breaks down and both the patient and provider are compromised.

How interpreters are selected and trained is the core issue in this section, however I will begin with an even more fundamental experience for everyone involved in this type of exchange: trust.

We all need to depend on other people at various times in our life. How comfortable we are with being dependent is a mixture of how confident we are with ourselves and what we think those other people may do with our vulnerability. Encountering a language we don't understand opens a range of choices and introduces the option of depending on an interpreter. Rosalind Edwards of London South Bank University, Claire Alexander from the London School of Economics, and Bogusia Temple of Central Lancashire University found "people who need and use interpreters are in a position of relative powerlessness…this sense of dependence and disadvantage will be an important factor in underpinning the relationship between individuals and interpreters, and in the articulation of trust." (Edwards, Alexander and Temple, 2006:11)

In their study, Edwards, Alexander and Temple offer the following insight from the work of Anthony Giddins (1984) on the concepts of 'personal' and 'abstract' trust.

> In relationships that are based on personal trust, comprising bonds such as family or friends, people are assuming that they can rely on these familiar others' consideration of, and stable commitment to, their own needs, interests and preferences. Abstract trust is called upon when a person is not familiar with another party but is reliant on them for their expert knowledge and competence. It is based on impersonal belief that a representative or member of a given expert group or institution will conventionally act according to particular principles, duties and requirements, which are laid down in codes and training for

professional groups, rather than their own personal interests.(ibid, italics added).

We all enter into relationships with professionals from various fields that we do not know personally but risk trusting based on our faith in the institutions which provide their credentials. The ability to begin a relationship with an interpreter based on knowledge of professional or government agency accreditation offers an external link to credibility. (see American Translators association (ATA) Chronicle Vol XXXV, 10:12 and XXXVI, I:31) On the more important personal side, between the interpreter and English-speaking client, professional credentials create a perception of distance between the interpreter and what might be viewed (suspected) as too much shared culture. For the person with whom the interpreter shares a cultural or community connection, credentials which are properly presented before the session, can provide reassurance of confidentiality and independence from the government agency. (Edwards, et. al. 2004) One of several potential conflicts in expectation arises when the agency or professional who uses an interpreter needs a feeling of trust based on the belief that the language transactions will occur minus any coaching from the interpreter and the refugee expects something quite opposite. In their study Using Interpreters to Access Services: User Views (2004) Edwards, Alexander and Temple report minority users define a good interpreter as someone who "does more then just change words into another language. People prefer proactive interpreters who can empathise with them, help them understand procedures and plead their case."

If information exchange is polarized with one extreme represented by transactional discourse for the purpose of conveying "factual or propositional information" and the other extreme as interactional discourse, described by Brown and Yule as "language to establish and maintain social relationships." (Brown and Yule 1983:1-3); interpretation might seem to fall into some middle area. From the refugee-user perspective, Alexander,

Edwards and Temple (2004, 4) reported, "Generally, however, the majority of people preferred drawing on their informal networks to act for them in all but important legal and medical matters. (Italics added.) The reason, as discussed in our earlier reference to trust (p. 10) "They trusted family and friends, rather than interpreters who were strangers to them, because they knew each other well and had a history of shared understandings and obligations." (ibid.)

Adding yet another layer to the complex expectations, some people in need of help do not want to share intimate details of their experiences with those from their social community for fear of embarrassment and a concern for privacy. This is a common reaction in cases of both male and female rape victims. The resolution to this situation involves awareness on the part of the health service provider and establishment of a code of ethical conduct for interpreters which will be discussed shortly.

As we progress through examples from different countries certain common themes emerge in terms of how interpreters are recruited, often on- the- spot with more emphasis on convenience than credentials. Where language brokers come from produced a pattern that has remained unchanged for centuries. Most people spend their entire life speaking to other people in the language they were born into. Absent war or natural disaster that would displace them into other countries, or choosing a profession that required contact with foreigners, learning a foreign language was a luxury for the elite or a means of survival for those on the other end of the economic scale. By any method of measurement or source of motivation, language services are a business and the providers are power brokers.

The terms used in one language to describe others who don't choose the same audible symbols reveals something about how we experience those 'others.' Slavic people, for example, referred to their Germanic neighbors as Niemetz, those without a tongue. The Greeks are credited with giving us the words bar-bar to describe sounds made by distant peoples, ignorant of the Greek language. The phrase is more recognizable in the spelling of Barbarian.

A common theme in the story of languages, and people who knew how to communicate in more than one of them, was the concept that masters don't learn the language of slaves. That may have a certain ring to it if you see yourself as a master; but to a slave, the same phrase represented a ticket out of their predicament. How people became translators and interpreters over the centuries, make that millennia if you consider the first known reference to a person designated as an interpreter was found in an ancient Egyptian hieroglyphic dating from 3000 BC, has remained somewhat constant in terms of drawing on communities of persons displaced from their homeland and birth language by choice, as emigrants or as a result of conflict, as refugees, prisoners or slaves. The fundamental requirement to speak at least two languages fluently is a common denominator for those taking the academic approach to language study or the refugee, emigrant or slave who must learn enough of the host country language to make a living, or just stay alive.

It was only recently, in the 20th century, that the United Nations established simultaneous interpreting as a permanent service available to its members in 1947. A new professional class, the conference interpreter, had begun to emerge at the 1927 International Labor Conference in Geneva, Switzerland. Twenty years later at the Nuremberg War Crimes trials, following World War II, the conference interpreter and a more complicated, audio technology dependent method, simultaneous interpretation gained recognition in the legal community. Still considered as much an art as a science, the simultaneous interpreter must hear one language and, with minimal delay, render the same meaning in a different language. Although sometimes whispered into the ear of one individual, the use of headphones and audio distribution systems has allowed many listeners to benefit from the language skills of one interpreter. (Burgener, 2005)

Even as the profession of interpreting and translating has grown and sought to define itself with standards for education and performance, the pressures to fill positions often turns language into just another commodity. Database searches of language use in the health care field confirm this approach is still very common.

For our purposes, general categories are defined as Lay interpreter: an untrained person such as a family member, often children. Ad hoc interpreter: a bilingual staff person who is called away from primary responsibilities – as noted in the South African study below, these persons may not be from the medical staff and don't always have an adequate vocabulary of terms. Professional interpreters include at the top end, conference and simultaneous interpreters and those employed in the court systems of several countries, for example who must be certified in command of a legal vocabulary. Demonstration of competence and command of medical, and particularly psychological vocabulary in not yet a reality.

Maria-Paz Beltran Avery, PhD of the U.S.-based National Council on Interpreting in Health Care describes the situation for people whose "immediate and frequently urgent health care needs do not wait for linguistic adjustment or cultural assimilation."

While the function of interpreting in order to allow communication between a patient and a health care provider who do not speak the same language has been going on for a long time, in the past this function was performed mostly on an ad hoc basis. Calling on whoever was immediately available – family members (including children), non-medical hospital staff, and even other patients – it was a practice that ran a high risk of inadequate communication resulting in misdiagnosis and inappropriate treatment that could, in a worst case scenario, result in the death of the patient. Such untrained individuals, often had little or no understanding of medical concepts or terminology and much less understanding of the importance of accuracy and completeness in the messages they conveyed. As a result, erroneous messages were often transmitted, new information added, or critical information omitted, drastically changing the nature of the original message. (2001, 2)

In her report, The Interpreter's World Tour an Environmental Scan of Standards of Practice for Interpreters, Marjory Bancroft (2005) notes:

In all, this scan reviewed 145 documents in 11 languages from 25 countries. Three dozen U.S. states were represented. Of these documents, 28 were called—or included—codes or guidelines for practice or standards of practice. No clear-cut distinction could be found between codes of ethics, codes of conduct, codes of professional responsibility, guidelines for practice, codes of practice, and standards of practice. Among the findings of the scan:

- Codes of ethics, conduct, or professional responsibility easily outnumbered guidelines or standards of practice for interpreters, by about 5 to 1.
- Documents were most commonly found in the industrialized nations with high levels of immigration: the U.S., Canada, Australia/New Zealand, and Europe.
- Few such documents were found in developing nations.

Bancroft goes on to describe the difference between a code of ethics and standards of practice:

The scan found that the vast majority of principles and requirements found in codes of ethics or conduct are also found in standards of practice, whereas the reverse is not so true. Documents about ethics or conduct serve to regulate interpreter behavior and address issues of "right and wrong," whereas standards of practice typically offer practical strategies to promote quality interpreting. Particular distinctions were noted:

- Professional standards for conference interpreting display a concern for logistics, contractual concerns, and working conditions.
- Standards for legal interpreters focus on reinforcing core ethics, in particular confidentiality, impartiality,

accuracy, and the need to follow the rules and regulations of the court.

- Standards for community and health care interpreters are often preoccupied with interpreter roles and boundaries, cultural mediation, client well being, and promoting client-provider relationships to ensure that the consumer's end needs are met

The issues of ethical conduct, particularly with respect to confidentiality, and the need to establish training and certification standards are recurring themes in reports on the role of language interpretation for refugees. (Temple, 2005)

The following is an example of guidelines for conduct of interpreters from the International Medical Interpreters Association Code of Ethics The IMIA Code of Ethics was established in 1987 and revised in 2006. Many of the elements outlined here appear in later discussions of how interpreters influence both the refugee and the health care provider.

I. Interpreters will maintain confidentiality in all assignment-related information.
II. Interpreters will select the language and mode of interpretation that most accurately conveys the content and spirit of the messages of their clients.
III. Interpreters will refrain from accepting an assignment when professional skills, family, or close personal relationships affect impartiality.
IV. Interpreters will not interject personal opinions nor counsel patients.
V. Interpreters will not engage in interpretations that relate to issues outside the provision of health care services unless qualified to do so.
VI. Interpreters will explain their roles and cultural differences or practices to health care providers and patients when appropriate.

VII. Interpreters will use skillful unobtrusive interventions so as not to interfere with the flow of communication in a triadic setting.

VIII. Interpreters will keep abreast of their evolving languages and medical terminology.

IX. Interpreters will participate in continuing education programs as available.

X. Interpreters will seek to maintain ties with relevant professional organizations in order to be up-to-date with the latest professional standards and protocols.

XI. Interpreters will refrain from using their position to gain favors from clients.

Does Anyone Know the Word For?

"Errors in medical interpretation are common, averaging 31 per clinical encounter, and omissions are the most frequent type."(Flores, et.al.2003,10) In their conclusion to a seven-month study involving 13 encounters with clients who did not share a common language with the health care providers, Glenn Flores, MD, Barry Zuckerman, MD, Milagros Abreu, MD, Leonardo Medina, MD and Eric J. Hardt, MD of the Center for the Advancement of Urban Children, Department of Pediatrics, Medical College of Wisconsin, Milwaukee, Wisconsin along with M. Barton Laws, PhD, Sandra J. Mayo, EdM, Department of Pediatrics Internal Medicine, Boston University School of Medicine, Boston, Massachusetts and the Latino Health Institute, Boston, Massachusetts went on to describe one of the key areas of vulnerability: the recruitment of individuals to serve as interpreters. "Most errors have potential clinical consequences, and those committed by ad hoc interpreters are significantly more likely to have potential clinical consequences than those committed by hospital interpreters." (op.cit. 9) However, the authors point out there was no statistical difference in their study between individuals designated professional hospital interpreters and those recruited on an ad hoc basis in terms of the mean number of errors committed per session. Although fewer in number, some of errors made by professional interpreters did have clinical consequences for the patients and that, the authors concluded, point to the need for better training.

In the US, less than 25% of hospitals provide any training for medical interpreters (only 14% offer training for volunteer interpreters) and in half of all the hospitals studied the training programs were not mandatory. (Ginsberg, et al.1955 in Flores, et. al, 2003, 10)

Dr. Flores and his team documented encounters involving children and parents with Limited English Proficiency (LEP) in a pediatric care department of a metropolitan hospital in the U.S. 'Clinical consequence' was defined according to the following examples:

Errors of clinical consequence included: 1) omitting questions about drug allergies; 2) omitting instructions on the dose, frequency, and duration of antibiotics and rehydration fluids; 3) adding that hydrocortisone cream must be applied to the entire body, instead of only to facial rash; 4) instructing a mother not to answer personal questions; 5) omitting that a child was already swabbed for a stool culture; and 6) instructing a mother to put amoxicillin in both ears for treatment of otitis media. (Flores, et.al.2003, 9)

The source of errors were identified as, "The most common error type was omission (52%), followed by false fluency (16%), substitution (13%), editorialization (10%), and addition (8%)." (ibid. 8).

At a psychiatric hospital in Cape Town, South Africa, researchers reported the following:

A significant proportion of patients are being assessed through the use of family members, cleaners and other inappropriate people. The financial burden to the hospitals of not providing an interpreter service is small, but the impact on working conditions and service to patients is considerable (Drennan, G. 1996,)

In the Swiss canton of Geneva, a study involving the review of over seven-hundred standard screening questionnaires administered to asylum seekers mainly from the Balkans and Africa between June and December 1998 reported the following:

The objective of this study was to determine whether language barriers during the screening interview affected the reporting of asylum seekers' health problems and their referral to further health care. Adequate language concordance was significantly associated with higher reporting of past experience of traumatic events and of severe psychological symptoms, contrasting with much fewer referrals to psychological care when language concordance was inadequate. These results suggest the importance of addressing language barriers in primary care centres in order to adequately detect and refer traumatized refugees. To address this problem, the use of professional interpreters is recommended. (Bischoff, Bovier, Isah, Francoise, Ariel and Louis 2003)

In their study, Bishoff, et.al. documented the difference in outcomes when "the subjective quality of communication was good or sufficient, symptom detection was significantly higher. When communication was rated as poor, the number of reported symptoms (physical and psychological) was two to three times lower." (ibid. 506). The difference between the use of trained interpreters and those simply recruited ad hoc was reflected in the type of symptoms reported. According to this study, interviews using ad hoc interpreters produced more symptoms of physical concerns (headache, abdominal pain and backache being the most common) but "significantly lower percentage of psychological symptoms." Psychological symptoms were most frequently described as insomnia, nervousness and sadness. (Bishoff, et.al. 2003,506,507).

In discussing their methodological issues, the authors of the Swiss asylum seeker retrospective study explained why they expanded the categories of language concordance to include one in which there was not complete understanding on the part of the nurse or the patient. This intermediate category is, unfortunately, where most of the medical interpreting falls based on my literature search. Bishoff and his colleagues stated:

We defined adequate concordance not only as both parties speaking the same language, but also as the two parties being able to speak their languages via professional interpreters. The intermediate category – partial concordance – was introduced when the nurse had only limited proficiency to speak a language understood by the asylum seeker or when a relative or friend acted as proxy interpreter." Italics added (2003,508)

The authors go on to describe one of the central differences between professional and ad hoc interpreters which we will explore in greater detail when we discuss how regulation and accreditation impacts the quality of language services.

To put the proxy interpreters automatically in partial concordance category may be considered as an arbitrary decision, but is based on the grounds of professional interpreters being trained, supervised, monitored and accredited while the proxy interpreters have no recognized language skills whatsoever, were not selected to act as interpreter by the nurse and did not necessarily translate in the main local language (French), but rather into other European languages including English, German and Italian. (ibid).

The issues of selection, supervision, monitoring and accreditation are fundamental to accountability for both the person serving as interpreter, the health care provider and the client's perception of privacy. (Bancroft 2005)

Much of the literature reviewed for this project over the past five years, began with a statement to the effect that there was very little written on the subject of interpretation in the field of health care. From my own observations, I concluded that what was available focused on the period when the refugees were in the process of waiting for asylum status in a host country with very little on the early phases while they were internally displaced persons or interred in relocation camps. With few exceptions most of the studies were prepared from the perspective of the host nation or international NGO medical providers and not the refugees themselves, unless the refugees were recruited as interpreters. In terms of the material available through a literature search, the lack of an accurate and comprehensive perspective from the refugees themselves neglects one portion of the already documented triad.

Writing in the Health Education Journal, P.J. Aspinall of the Centre for Health Services Studies, University of Kent observed, "ethnic inequalities in health and healthcare are marked and persistent and language has been identified as a key barrier to accessing services and effective communication. (Aspinall, 2007, 90). The studies outlined in Table 1 provide a sample of the contemporary efforts made by the medical service providers and refugee agencies; Aspinall states the National Health Service (NHS) in the UK uses language data that is "10-15 years out of date" for its Resource Allocation Weighted Capitation Formula, a statistical method for deciding where funding will be awarded. According to Aspinall, the upcoming 2011 Census offers an important opportunity to gather comprehensive data.

As outlined in the first pages of this paper, there appears to be a progression from instincts to desires and then demands. In terms of language, who is exactly in a position to make demands? In Immanuel Kant's essay Perpetual Peace (1795) he sets forth a "Third Definitive Article for a Perpetual Peace" in which "The Law of World Citizenship Shall Be Limited to Conditions of Universal Hospitality." Kant goes on to describe those conditions in detail including the "right of a stranger not to be treated as an enemy when he arrives in the land of another," and notes the extension of hospitality and the withholding of hostility "is not the right to be a permanent visitor that one may demand." It is, Kant argues, "only a right of temporary sojourn."

The 20th Century myth of the melting pot where the pot was an environment defined by a majority who extended the hospitality of being melted to minority newcomers is being replaced by a different paradigm. The minority no longer feels compelled to accept the hospitality of melting. Identity is defined by the individual, not geography or politics.

Instead one must build a new sense of national (state) identity in which citizenship is not based on allegiance to an indivisible community, but it is based on a contractual arrangement in which the citizen agrees to respect and tolerate others without requesting a commitment to one cultural tradition (or ethnicity). (Nimni, 2007, 7)

Where does language fit into this perspective? In 2004 then British Home Secretary David Blunkett told BBC radio: "We have norms of acceptability and those who come into our home – for that is what it is – should accept those norms just as we would have to if we went elsewhere." (Emphasis added). (The Guardian, 27/2/2004 in Alexander, Edwards and Temple 2006, 3). The British politician is not alone in his view of what constitutes good manners on the part of guests. While the specific language changes from host country to country, the idea of 'foreigners" learning the language of the natives is a recurring theme in immigration debates. The Heritage Foundation, a think tank representing the views of many conservative Americans recently published the following account. "With the most expansive immigration and naturalization overhaul in the past 40 years languishing in Congress, it is worth pausing to reflect on the wise words of Alexis de Tocqueville: 'The tie of language is perhaps the strongest and the most durable that can unite mankind.'" The authors, Matthew Spalding, Ph.D. and Israel Ortega continue, "Critical to that success has been the role of a unifying and singular language. The ability to converse, interact, trade and communicate in a common language is key in order for newcomers to assimilate into the nation's unique fabric and become active participants in - and valuable contributors to - society." (Spalding, Ortega, 2007)

In a similar example of what can be seen as cultural arrogance, the British Home Secretary wrote:

We must strive to connect people from different backgrounds, tackle segregation and overcome mutual hostility and ignorance. Of course one factor in this is the ability of new migrants to speak English – otherwise they cannot get good jobs, or share in wider social debate... For those long settled in the UK... speaking English enables parents to... participate in wider modern culture (quoted in The Telegraph, 17/9/02, emphasis added in Alexander, Edwards and Temple 2006, 4)

But is it really arrogance when people take pride in their heritage and the language that is part of that experience? Is it unreasonable to want to hear familiar, recognizable sounds? No, I think it is not unreasonable; however people must be prepared to accept that others may well hold the same feelings. When those feelings are expressed as demands or "rights" then an inherently personal matter becomes a politicized issue. Variations of this theme are referenced throughout this paper and were articulated in an example from Alexander, Edwards and Temple in their 2006 essay Contesting Cultural Communities: language, ethnicity and citizenship:

The role and status of the English language exemplifies an unresolved tension between citizenship and cultural, ethnic, racial and religious diversity. To acquire language is to do more than acquire the ability to communicate, it is to acquire culture (Bhabha 1994, Fanon 1967). The culturing of language is especially problematic when it is mapped onto a vision of national identity, with its concomitant ideas of citizenship and belonging, insiders and outsiders and the implementation of differential rights and obligations (for example, access to welfare services (Lewis 2000).

Domestic and international law provide an expression of the values we assign to certain aspects of our lives, the 'differential rights and obligations,' to the point of providing a system for protecting them. If language has a value then the law should offer examples of how it is protected. In terms of guarantees of the right to interpreters, the documentation is less specific.

In article 14 of the International Covenant on Civil and Political Rights adopted and opened for signature, ratification and accession by United Nations General Assembly resolution 2200A (XXI) of 16 December 1966 with entry into force 23 March 1976, paragraph 3 states:

In the determination of any criminal charge against him, everyone shall be entitled to the following minimum guarantees, in full equality: (a) To be informed promptly and in detail in a language which he understands of the nature and cause of the charge against him;

Writing for the International Journal on Minority and Group Rights, Henrard explains:

Language use in court proceedings, in prisons, and in communications with the police are three areas of interest within the domain of the administration of justice. If the existing standards that are of explicit or implicit relevance for language use in the administration of justice are analyzed, the relevant general human rights provisions, minority rights standards and prison rules should be studied. Whereas the first category of rights reflects concerns of justice and fairness, and are thus formulated in terms of language one understands , the second one refers more directly to minority languages (and thus also languages of the various population groups in plural societies without majority population).

However, the latter provisions are so vague and replete with escape clauses that their actual contribution to the accommodation of linguistic diversity in respect of the administration of justice remains doubtful....Overall, it is obvious that this area of law and its specific problems are little analyzed and not yet fully explored. (Henrard, 2000)

Article 8 of the World Health Organization Declaration of Cooperation published in Geneva in January 2001, concludes with "Access to interpretation should be guaranteed when refugees and displaced persons are dealing with authorities, various services, or agencies. (Petevi, ed. 2001)

Access to an interpreter is not a legal right except in criminal proceedings. That has not prevented the leveraging of Civil Rights law in many countries to accomplish the purpose of providing language services.(Mikkelson, 2000; Bancroft, 2003)

The use of the term role may evoke a theatrical metaphor if we are able to see language as one of an ensemble of players engaged in acting out a slice-of-life drama. Herbert Read (1951 in Klein, Heiman and Money-Kyrle 1955,83) reminds us, "The first perceptions of what is novel in any science tend to assume the form of metaphors – the first stages of science are poetic." The behind the scenes motivation of each actor is rooted in Klein's description of the epistemological need to create order, displacing feelings of impotence in the face an overwhelming onslaught of unanswered questions. Control might be recognized through Edgcumbe's description of Vigotsky's intermental and intramental stages of language acquisition and established through mastery of Dollard and Miller's cue-producing responses in the form of reasoning and planning which can yield the all important and stabilizing influence of predictability. Producers and directors, those responsible for proving the material means and continuity needed to create an image or identity for the performance in the mind of the audience are, in the case of language and interpreters, the administrative and political forces controlling funding and regulation.

The fact that language is only one of the factors impacting anyone's access to services does not diminish the importance of the role it plays in the refugee experience or how people from other language groups experience refugees. Because language is an easily observed discriminator it is often the starting point for discussions of problems and solutions. There is a superficial dimension to any debate based solely on language because, as I have outlined in the theoretical structure, our verbal behavior reflects a wide range of conscious and unconscious emotions. Refugees and the people who work with them in host countries represent a cross-section of humanity including good people who will continue to behave in an ethical and morally responsible way despite their circumstances of power or powerlessness and those who will exploit any situation to their own advantage.

The examples presented in this paper were drawn from a diverse geographic sample ranging from Norway and Switzerland to South Africa, the United Kingdom and the United States involving an equally broad range of languages and cultures. The common denominators in each situation are human beings, influenced by the factors of personal perception of interaction with external objects described in our opening introduction to theoretical structures. Expanding on that theoretical base and the political discourse surrounding issues of cultural and ethnic identity and reaching well beyond any metaphor are observed facts such as these:

- No interpreter was available for the parents of a 3-year-old brought to the emergency department because of abdominal pain; she was discharged twice, returned with an acute abdomen, and was hospitalized for treatment of appendicle rupture and peritonitis.
- A 2-year-old fractured her clavicle in a fall but was placed in social services' custody because of a pediatrician's misinterpretation.
- Parents of a ventilation-dependent 2-week-old with encephalopathy, seizures, and renal failure were unaware of the infant's poor prognosis, despite use of an interpreter. (Flores G, Abreu M, Schwartz I, Hill M 1999, 1253)

No discussion of the role of language is accurate, much less complete, without including the clinical impact on human life resulting from errors in interpretation. The demonstrated source of many of those errors is the poor choice of interpreters. If better outcomes are desired then informed choices need to be made and accepted within the triadic relationship.

While the motivation to improve language services should come from moral and ethical foundations, it is more likely to find its impetus within the legal framework of medical malpractice lawsuits. If the risks associated with a procedure are explained to a patient in advance for the purpose of allowing the patient to make an informed decision, then it can be argued that failure to provide information in a language that is understood by the patient invalidates their consent. A study by the Center for Ethics and Humanities in the Life Sciences at Michigan State University posed the question: "Are Good Intentions Good Enough?: Informed Consent Without Trained Interpreters." It notes, "Some states have adopted Medicad codes that include compensation for language interpretation. Clinicians, who may be legally liable for failure to assure adequate interpretation in their clinics could more avidly pursue such changes." (Hunt, de Voogd 2007;604, italics added)

Establishing Standards for Interpretation

If the role of language interpretation is taken on by actors with neither talent nor technique the audience suffers through each performance and in the case of accessing medical care, the reviews may take the form of obituaries.

In Court Interpreting at a Crossroads, Holly Mikkelson of the Monterey Institute of International Studies provides the following insight, "No formal training in court interpreting was offered until government entities began setting proficiency standards for interpreters in the courtroom." (Mikkelson, 2000)

The certification movement in the U.S. has been confined to judiciary interpreting, at least for spoken-language interpreters; in Australia and Canada, in contrast, a multi-tiered program of accreditation has been established. Professional interpreters in Australia, for example, are rated at four different levels: 1) Paraprofessional Interpreter, 2) Interpreter, 3) Conference Interpreter, and 4) Conference Interpreter (Senior). It is recommended, though not required by law, that court proceedings be interpreted by interpreters at the third level or above; those who qualify as Interpreters are tested only in consecutive interpretation (NAATI, 1999). Another singular characteristic of the United States is that the regulation of court interpreting was initiated by the legislature at the federal and state levels, albeit with input from professional interpreters. In most other countries, the impetus has come from professional associations, and the exams are administered either by the associations themselves (as is the case in Canada and the U.K.) or by an independent body that works closely with the professional association (as in Australia).

In March 2007 the Centre for Analysis of Risk and Regulation (CARR) at the London School of Economics hosted a conference under the umbrella heading of "Organizing risk regulation: current dilemmas, future directions." Although public agency provision of language services was not on the agenda for discussion, parallels could be drawn with existing regulated professions. In his opening remarks, Professor Peter Miller, CARR Theme Director, stated:

> The British tradition of the self-regulating profession increasingly appears a thing of the past. Professions including law, healthcare, and accountancy face new regulatory institutions, increasingly pervasive modes of regulation and oversight by lay people. All this takes place against a background of increased consumerism, a perceived decline in public trust, and pressure on regulators to risk-base, target and reduce burdens. (CARR Conference Proceedings, 2007)

If regulation, as in the case of court interpreters, leads to higher standards and better performance then who will take the lead in demanding regulation of interpreters in the healthcare field?

In the access to and delivery of health services, poor interpretation leads to poor results; Alexander,C. Edwards, R. and Temple, B. with Kanani, U. Liu, Z. and Miah, M. (2004), Alexander,C. Edwards, R. and Temple, B. (2006), Bancroft, M. (2005), Bischoff A, Bovier PA, Rrustemi I,Gariazzo F, Eytan A,Loutan L. (2003). Poor interpretation is a statistically significant outcome of the choice of interpreters; Flores, G., Laws, M. B., Mayo, S. J., Zuckerman, B., Abreu, M. Medina, L., MD and Hardt, E., (2003), Flores, G. MD, (2005), Bischoff A, Bovier PA, Rrustemi I,Gariazzo F, Eytan A., Loutan L. (2003) Jones, D. (2007). Formal regulation of language services in terms of required training, certification and monitoring offers a viable alternative to the present situation in the health services field based on the experience of court interpreters; Mikkelson (2000), and the conclusions articulated by many of the studies cited here including: Bancroft (2005)

Language and particularly language which must be interpreted to be understood plays a major role in a refugee's ability to access services anywhere that contact with something other than their birth language occurs.

Limitations of this Study

The choice of what we considered to be "appropriate" examples followed many of the descriptions contained in Jennifer Hunt's Psychoanalytic Aspects of Fieldwork including this: 'There are few researchers whose object of study does not condense a richly colored prism of transferences.' (1983:30). My own past experiences with refugees and language influenced the choices, however the database searches themselves contained only an assumption that errors were made during interpretation.

The commingling of refugees with other users of foreign language represents the reality of how language is observed and reported as a factor in access to services. This does not diminish the impact of language on refugees and in fact reinforces the role of language as a determinant of how successful anyone can hope to be in this particular arena of human interaction.

Suggestions for Further Studies

In 2005 Glen Flores, MD and a team from the University of Wisconsin Medical School reported:

> A systematic literature review was conducted on the impact of interpreter services on quality of care. Five database searches yielded 2,640 citations and a final database of 36 articles, after applying exclusion criteria. Multiple studies document that quality of care is compromised when LEP patients need but do not get interpreters. LEP patients' quality of care is inferior, and more interpreter errors occur with untrained ad hoc interpreters. Inadequate interpreter services can have serious consequences for patients with mental disorders. Trained professional interpreters and bilingual health care providers positively affect LEP patients' satisfaction, quality of care, and outcomes. Evidence suggests that optimal communication, patient satisfaction, and outcomes and the fewest interpreter errors occur when LEP patients have access to trained professional interpreters or bilingual providers. Key Words: translating communication barriers language physician-patient relations quality of health care patient satisfaction. (Flores 2005, 255)

Improving access to trained interpreters is an issue requiring further exploration.

With this and similar studies available, David Jones, a general practitioner in the UK writing for the British Medical

Journal in 2007 felt compelled to question: "Should the NHS curb spending on translation services?" Reflecting on his own experiences with multiple generations in Tottenham which he describes as "a deprived inner city area of London with a diverse population of new migrants" he argues for more spending on language services. He concludes, "We must not let the politicians persuade us that it is the patients' fault." (2007,335)

What is needed then is not simply more studies from a healthcare or linguistic perspective but a public policy debate on the broader issues which are subject to political resolution.

The application of computer-based technology has yet to achieve its full potential, however, as the following Norwegian study suggests, it is an appropriate alternative.

In the health care system, good communication is of vital importance for correct information, diagnosis and treatment. When the health care worker and the patient have no common language, an interpreter is needed. In a small population the patient and the interpreter will often be acquainted. In small, local communities few professional interpreters are available. These communities can be served by offering a telephone interpreter, thus providing foreign citizens with a better health service. The presence of an interpreter reduces the possibility of being anonymous. The patient may withhold important information, or give incorrect information. By using a telephone interpreter, neither the patient nor the interpreter needs to know who the other person is. We found this to be a very good alternative in most cases, and sometimes a better solution. The interpreters were not as satisfied as the doctors and patients. Further development of the service is therefore required. (Karlsen W.B, Haabeth A.L.1998, 253)

Further progress will be made through use of technology to manage the demand for language services and psychoanalytic concepts to understand our resistance to knowing.

Chapter 6

Emergency Medicine Environments

As an emergency physician over many years in multi-cultural communities I have frequently encountered many patients who do not understand or at least do not comfortably speak English. This presents a very big challenge. The most critical parts of good medical care are the history, physical exam and well understood treatment including follow-up care.

One can usually get through the physical exam with an assortment of expressions, hand signals and descriptive sounds and pantomimes. But the rest requires the use of language and appropriate interpretation. Acute injuries such as burns, lacerations, fractures, sprains are relatively easy because the condition is obvious and can be treated in an obvious and concrete way and follow-up instructions are usually simple and straight forward. Chronic conditions such as diabetes, hypertension, coronary artery disease are more difficult. The symptoms, diagnosis, treatment and often necessary life style changes are more subtle with more lasting effects.

With a medical history, are we even talking about the same things? People have such different experiences with medicine and treatments and also have definite cultural differences. Other cultures tend to take care of their own with improvised care and home remedies first. Their access to doctors and medical care is often limited. They think Americans are soft and pay too much attention to health. They feel it is not manly or macho to admit sickness or weakness. They often do not go for health care until they are very sick or the cancer has metastasized or the peripheral vascular disease has become gangrene. Many American groups also react this way - it is not only "they". When it comes to immunizations, did they have them or not? When they say yes, does this mean they really did or do they say yes because they think that is what we want them to say?

Do they have a support system really? Many do; they are part of a language and cultural group who stick together and are usually supportive. But they also band together and hide those who are sick or need help and are unable to get it for one reason or another. Few have insurance and it is probably not adequate. Many have poor English communication, have no jobs, or have jobs but no money, or are illegal immigrants. Some have communicable diseases such as sexually transmitted diseases , tuberculosis, or HIV-AIDS and are afraid to admit it.

An attempted solution to the dilemma has been to push for the use of certified interpreters. This of course presents the question of advantages and disadvantages of "certified interpreters" versus family/clan members and/or interested persons. How well do foreign speakers open up to total strangers even if they speak their language? If they come from a police state do they feel they are being interrogated, threatened, or coerced to give up more, violate taboos, and become even more vulnerable?

We have been fortunate to have access to some certified interpreters who functioned in such a way that was very efficient and all involved felt it led to a satisfactory outcome. I recall one Russian interpreter whom I had never seen before and a Korean interpreter who is very gracious and always comes in when he is needed. With both, there was a discussion with me regarding the medical condition, with best treatment and follow-up after a discussion with the patient and family regarding the history and symptoms. That is always followed by my asking if the patient or family have questions. In both cases we were able to contact the patient's own doctors who spoke with me in English and also with the patient and interpreters in their own languages. These incidents represent not only optimum care but also the best that can be expected.

We also have a Portuguese speaking woman who is not certified but is studying for certification. She is part of a Portuguese speaking community and frequently brings other members in for treatment. We have a good relationship with her. She is very helpful about explaining some of the cultural treatments and attitudes. One truly memorable example is the Brazilian worker she brought in with a significant burn of his hand which he had covered in coffee grounds. Amazingly he had no pain until we cleaned off his hand and administered the standard American burn treatment. I had never seen this before and questioned the patient extensively through the interpreter. She explained that in many parts of Brazil people both live and work in areas which are very far from any medical personnel. Many of these areas are primitive camps in the jungle. They all have campfires and many have primitive metal stoves and burns occur frequently. In this part of Brazil as in many parts of the world water is not potable, but coffee is a mainstay in most diets. When significant burns occur they are treated with the most available material and that is the already brewed coffee grounds. Our patient stated that in Brazil they usually covered the coffee grounds with honey but since he didn't have any honey he used a cloth instead.

I haven't found anyone yet to explain this phenomenon. I am assuming that the physical barrier forms between the burned tissue and the environment where any difference in temperature or air current can trigger a pain response in the exposed nerve ends. I am also assuming that the barrier is formed by any material which is readily available - much as butter or lard was used on burns when I was growing up in the Midwest. Whether there is some other component in the coffee grounds themselves which potentiates the effect is unknown to me at this time.

She also is very helpful about arranging for follow-up with a doctor who speaks their language and will see them even if they have no insurance and will help them at a pharmacy for their medications.

We have also had certified interpreters who were less than helpful. The have listened to lengthy questions or descriptions from the medical persons and given an 'interpretation ' to the patient of a few words only. Then they have turned a lengthy complaint from the patient to the doctor with a few words only. One must question the adequacy of this relationship and it is more frustrating than helpful.

We had one episode where we had a lengthy search for a Lithuanian speaker who came in and interviewed the patient (who was a very old lady). He could not understand her and thought it must be some obscure dialect. We finally found a Lithuanian/English speaking relative who said the family had not been able to understand her for years as she was demented. They finally brought her in desperation because they thought since we were doctors we could figure out what was wrong with her. Then we periodically see people who bring in a relative who has supposedly lived with them "for a long time" but they cannot answer any questions about the patient, the complaint or the history.

One other situation has arisen lately which creates concern about our reactions. We are seeing many more Mid-Eastern/ Arab/ Muslim patients with the women covered head-to-toe walking behind their husbands and with very different customs. Do they think we are criticizing them because of the difference and because of the societal attitudes and prejudices? Underneath the attempts to understand and care for them in the same way we attempt to understand and care for anyone else is the lurking knowledge that they are different although maybe only in dress, accent and customs. But is that the difference or is there something sinister and dangerous about that difference?

There seems to be a heightened level of the usual situational awareness where anyone - in ordinary or in different dress - could be a serious threat. This is something which requires further understanding. Emergency rooms can be very dangerous places. We do not want our reactions to create problems. But neither can we neglect the fact that threats do exist and without understanding and situational awareness, damaging incidents do occur.

In 2001 while I was still in the U.S. Army Reserve Medical Corps I was mobilized to Kosovo/Macedonia in support of the NATO mission in Kosovo. I was assigned to the "rear supply base" in Macedonia where the supplies which had come in by sea and transported by ground to us were reconfigured and convoyed by ground to the main NATO bases in Kosovo. During the time of my assignment there an Albanian led insurgency was building against the Macedonian government and there were military confrontations all around our camp. The rebels took a small village at the foot of the mountains and from there launched mortar attacks across one of the main highways into Skopje - the capital. This was just a few kilometers down the highwayfrom our camp - so close that when the helicopter gunships launched their counterattacks they hovered right over our camp when they fired the rockets into the village. After the bombardment and submission of the village our (U.S.) forces were tasked with moving the rebels back into the mountains away from the capital. Our medics and ambulances were assigned to bring any rebels who were significantly wounded back to our aid station/clinic on the base for medical care and for transfer to the combat support hospital in Kosovo if needed. We received four casualties with very significant wounds including one with partially collapsed lungs and another with one leg fractured. In addition all of them had developing wound infections as we could not get them out until a couple of days after the battle when negotiations could be completed through diplomatic channels. One of the surgeons and his team had been sent down from Kosovo just before the bombardment started and they were available to help with the casualties. They took two of the patients in the other room. I was told the patients were refusing IVs and medications. We had a native Macedonian who was fluent in English and had been assigned to our clinic for several months. He told me later that one of the prisoners was from his hometown. The group spoke Albanian which none of us spoke. Most of them spoke Macedonian so our interpreter was able to talk with them. One of them also spoke some German and I think understood some English. He heard me ask the interpreter to find out how many days since they were wounded and before the interpreter spoke, he replied "zwei" which of course is German for two. The man with the fractured leg was older, appeared to be observing his surroundings very carefully and seemed to be the leader of the group. I told the soldiers to keep an extra eye on him. Even though he was significantly wounded, I could not completely trust him. He was not trusting me either. I had wanted to have the medics insert an IV so we could get him better hydrated, give him some antibiotics and pain medication, but he refused. He did ask for a drink of water. I asked our interpreter to explain again that he would be needing surgery and until the surgeon had seen him I couldn't give him anything by mouth and that was why I had wanted to start the IVs.

Then our interpreter said "Ma'am, he didn't actually refuse an IV - they just did not understand each other in the other room." I then asked if they knew what was happening to them and he answered that they knew they were being transferred but were concerned about where. Suddenly I realized they might be afraid we would turn them over to the Macedonians. I then asked our interpreter to explain that they were to be sent in a helicopter to the American hospital at Camp Bondsteel - the same as we do for our soldiers and they would not be taken to the Macedonian hospitals. I could see the relief in his face as he understood and then explained in Albanian to the other patient who also looked relieved. I asked our interpreter to go in the other room and also explain to the other two patients. As he left the room he stated that the leader would now take the IV and other things we needed to do for him. From that moment until the helicopter lifted off with them on board we had complete cooperation from them."

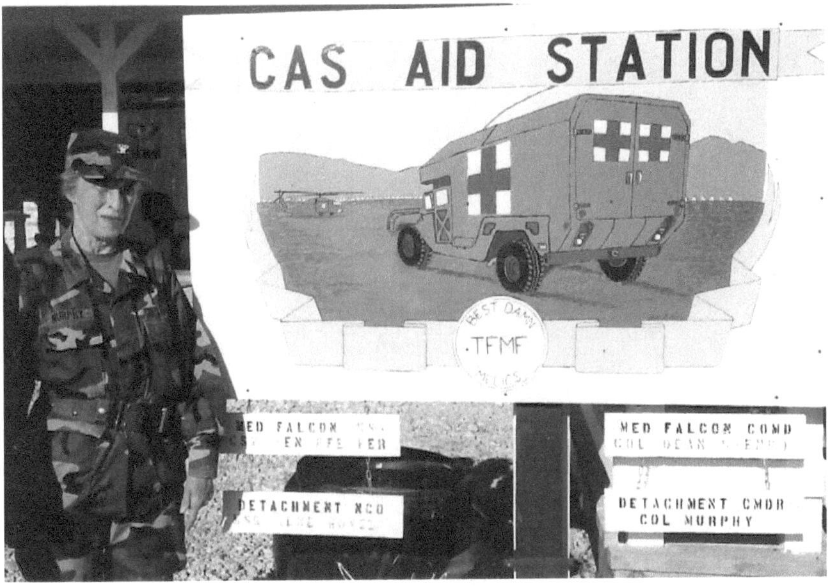

Chapter 7

Where's the Language Crash Cart?

Wasting time can do as much damage to a patient as the initial injury or illness. Matching the correct language skill sets of an interpreter to the patient who was just carried in must start well before the patient ever arrives, rather like re-stocking the crash cart before it is needed. Then there is the reality of where that particular interpreter is when you need them- are they with another patient? Is today their day off? Closing the distance in a timely manner can be accomplished with something as simple as a telephone. How various technologies can be applied to the issue of bridging language differences will be presented in detail with the next book in this series, *Connecting Babel.* Because timing is important we will share some thoughts as they pertain to language issues in emergency medical situations today even as progress is being made.

According to the International Medical Interpreters Association, medical interpreting is the fastest growing segment of the language service market. Unfortunately it is never fast enough and the methods used by NATO countries and the UN to recruit interpreters during a conflict don't hold out much hope that the medical vocabulary would be where it is needed. As we have already discussed military operations provide examples of how the latest in medical technology can fail in the mission of saving a human life when a simple exchange of information can't be readily established. When civilians are injured, military medics are trained to handle the bleeding and broken bones and stabilize patients for transport to civilian hospitals, if available, or to a military health care facility. On the asymmetrical urban battlefield where most fighting takes place today, the range of medical issues reflects the full gambit of what can go wrong with human bodies – not just the sucking chest wounds medics are trained to deal with. In the Balkans and Iraq, where Dr. Murphy and I have firsthand experience, to probably most other conflicts of the past decade, ordinary soldiers and marines are being asked to manage everything from old-age issues of dementia and Alzheimer's to mental retardation, hereditary deafness and speech problems. The two key factors involve medical conditions not normally found in the ranks of healthy, young combatants and exposure to language dependency at all levels from those military personnel involved in the first contact up through the various levels of detainee or prisoner handling.

In January 2004 part of the infamous 240 acre Abu Ghraib complex near Baghdad had been re-designated the Joint Debriefing Interrogation Center (JDIC). As one of the officers in charge of screening operations my team was responsible for reviewing capture documents and conducting initial interviews with people who had been "detained" by coalition forces. One batch of new detainees included a young man in his early twenties who didn't respond to verbal commands in any language. Was he a foreign fighter or simply someone who had been trained in how to evade interrogation? It took some doing but we finally established that he was deaf due to a congenital hearing problem. Our Air Force medics were well trained and equipped to comply with Geneva Convention requirements but that did not include hearing exams.CW3 R.D.Burgener

---

If the person you are trying to help is conscious the range of options for establishing what language they speak is much greater than if you must rely on identification documents, if they have any, or other printed material they may have on their person. What you can bring to the process is a poster or small laminated card with the flags of countries and the name of the country in your language and the language of that country. Do you need to do the entire world? Start with the languages and countries you have dealt with in the past and can feel reasonably sure are present in the demographic mix of the community you are serving.

Izabel Arocha, M.Ed. Cultural & Linguistic Educator Cambridge Health Alliance and president of the International Medical Interpreters Association offers the following suggestions; "There are two other ways in addition to flags, which usually don't correspond with languages, especially in a country like India where there are numerous dialects. There are 'call an interpreter' cards, where an interpreter identifies the language in the back of a card, or simply 'I speak' cards, which the patient uses, the size of a business card. There are also 30+ language posters alerting patients of their right to an interpreter, in many languages, asking them to point to the language. Last, there is a great language identification service by Language Line Services, where linguists who are trained to identify languages (Defense training) identify the dialect over the phone. There is also a great website www.ethnologue.com which is used sometimes by hospitals when trying to assess which specific dialect the patient is speaking."

The organized noise we recognize as language (even if we may not understand) is a major component of human communication. It is not, however, the only way humans can exchange information. From the perspective of personal identity, not communicating with sound or according to what the majority might consider an organized sequence, doesn't make a person any less human. It also doesn't make them expendable during life-threatening emergencies. Telling people caught up in a disaster that you care about them and want to help doesn't carry much credibility unless your language crash cart is stocked with communication tools that get your message across and give them a means to respond.

Amy Wilson-Stronks, MPP, CPHQ, Project Director, Division of Standards and Survey Methods Principal Investigator, Hospitals, Language, and Culture Study for the Joint Commission, described circumstances other than verbal language which can inhibit communication.

- Hearing Impairment
- Visual Impairment
- Cognitive Limitation (Autism, Fragile X Syndrome, other forms of mental retardation)
- Intubation (when a breathing tube or other device is inserted in the victim's mouth as a life saving step. This usually makes normal speech difficult to understand)
- Disease (ALS, Stroke)
- Health Proxy (when the person with the victims power of attorney to make medical decisions, or parent of a minor is not available)
- Culture
- Literacy

Our goal in this book is to provide examples of how language defines difference when people attempt to communicate. It is worth noting that language has more than one form.

Susan F. Rzucidlo, founder of SPEAK Unlimited Inc. works with special needs communities and First Responders in the state of Pennsylvania. According to Rzucidlo, "The Premise Alert System allows families to voluntarily notify the police and other first responders about their special circumstances free of any charge. It may be a child or adult with autism, developmental delays, or mental retardation, who is known to wander. It may be someone with rare or complex medical issues that require special handling. It could be an Alzheimer patient who wanders. The Premise Alert System also helps police departments identify individuals who have special needs, which will enable the responding officer to have additional information at his/her disposal. This form will be distributed to families through case managers at the Department of Mental Health and Mental Retardation as well as schools, community groups and legislators' offices." Visit their website for more information: www.papremisealert.com

Known as the Summit/Loraine Project, Sheriff Drew Alexander of Summit Ohio and Chief Celestino Rivera of the City of Loraine teamed up with Isabel Framer an Oregon/Tennessee Certified Court Interpreter (Spanish), now living in Copley, Ohio to produce a "*Flash Card Communications Booklet*" to help police and victims determine which language is needed and communicate until an interpreter arrives.

In 2003 Framer began a nationwide search for law enforcement policies with well articulated guidance for dealing with LEP situations. " She found that most policies were not in writing or were incomplete." (Summit/Loraine Project, 12) We can all hope that is changing – but just hoping won't get it done.

Chapter 8

Trust in the Language Triad

Trusting people, in its simplest form, means we believe we know how another person is going to behave. Confidence is the result of our trust being validated by the other person's behavior. As we all know, sometimes that just doesn't happen and so we must contend with the element of risk. Building trust is something of an exercise in risk management and success begins by understanding what might be going on in the mind of each participant, including yourself.

Interpreters are people with skills in more than one language and culture. How you have integrated those experiences is part of who you are as a person. The people who use your skills, however, may feel threatened because they can only identify with part of you – the part that speaks their language. That would be difficult enough, but some human beings have an unfortunate capacity to become paranoid about things they don't understand, things that aren't predictable. So while you are doing your job and speaking another language, at least one of the three elements in this language triad may be feeling uncomfortable. When people exchange information directly, for example one-on-one or a speaker to an audience, the communication is *dyadic* because there are only two directions for the information to travel. When people cannot communicate directly with each other because they do not share a commonly understood language, then a third person who understands both languages must be inserted into the mix creating a *triad.*

Participation in a conversation taking place between people who share a common language can be and usually is colored by the various dynamics of personality, history and all the emotional baggage we bring to human interactions. Even if the people and conversation are divided along religious, racial or political lines with an ensuing shouting match pushing emotional limits; one of the emotions usually absent is a fear of not understanding the language. That doesn't mean you won't hear people say, "I don't understand you," but in this situation the reference is to content or motivation, not the words themselves. When a second or foreign language is used, the fear of not understanding along with uncertainty about trusting someone who speaks both languages becomes a player in the competition for power.

Rosalind Edwards of London South Bank University, Claire Alexander from the London School of Economics, and Bogusia Temple of Central Lancashire University found "people who need and use interpreters are in a position of relative powerlessness…this sense of dependence and disadvantage will be an important factor in underpinning the relationship between individuals and interpreters, and in the articulation of trust." (*Sociological Research online,* 2006:11,1)

How well those of us who need language services manage our comfort level makes a big difference in the outcome. As an interpreter, you can help us, and yourself, by simply being aware, particularly with inexperienced clients or your first time with a new client, that you are seen as a broker not only of words, but of power.

The captions for the illustrations that are an integral part of this essay are direct quotes or at the very least accurate paraphrase of comments made by members of each segment of the triad. In the context of formal contractual relationships or simply real world situations, the segments are created along the lines of two people who want or need to communicate with each other but can't because of language differences and a third person who can resolve those differences and create communication. The interpreters role as the third person involves working for a client who contacts and retains their services and another person whom we will call the "client's client."

The first group is 'clients,' in my experience were soldiers from many different countries serving with coalition forces in the Balkans and Iraq. After I retired from the U.S. Army and returned to post-graduate studies in England I experienced many similar comments from social workers and health care providers. Let's begin with the client, military or civilian, who believes he or she has bought and paid for the right to be able to trust you. Misinterpretation then in not just a violation of trust, it is a breach of contract. Their mantra is "The Interpreter Works for ME!"

The mind set of expecting trust by virtue of financial obligation is often a factor anytime people with no experience in using a language other than their own, or working in a foreign environment, must put all their anxiety in one container labeled: interpreter or translator. A language dynamic on at least two levels has been created as U.S. and European and Asian military assets are extended into conflicts from Africa to the Balkans and Iraq to Afghanistan. First, as you well know, there just aren't enough of you trained and certified interpreters to go around. The Iraq Study Group Report (2006 p.92) states: "All of our efforts in Iraq, military and civilian, are handicapped by American's lack of language and cultural understanding.. Our embassy of 1,000 has 33 Arabic speakers, just six of whom are at the level of fluency." The U.S. Department of Defense, for example, approached their need for Arabic, Pashto, Urdu and other languages by advertising in target language newspapers circulated in the emigrant communities within the U.S. and ad-hoc recruitment of "interpreters" in the countries where our troops are deployed.

Many who answered the call for the War on Terror have been used up and discarded at the end of their contracts with little or no attention paid to their mental health needs. The concept of Secondary Post Traumatic Stress has been documented among journalists covering wars and to a lesser extent among interpreters who must ask questions and repeat answers involving unspeakable experiences. In the context of risk management, interpreters and translators may have valid concerns about trusting their employers. A story in the February 29, 2008 edition of the Washington Post began with the headline "Visas for War Zone Translators Halted Current Quotas are Nearly Filled" The story describes the halt as "the latest obstacle for many of the several thousand translators who have worked for U.S. military units in Iraq and Afghanistan, risking their lives and leaving their families vulnerable to retaliation from insurgents who see them as accomplices of American troops." It was reported that many American service members had acted as sponsors for their former translators, testifying to the bond of trust that can be formed. That takes us to the second level of the language dependency dynamic.

The nature of conflict today involves contact between soldiers of all ranks, most of whom have no training or experience in working with foreign languages, and civilians who may or may not be combatants. Sorting out friend from foe with words and not bullets often comes down to which you have more of - and which one you trust. Trust between individuals can be built only to be torn down and betrayed by distant decision makers with neither the experience of risk nor the ethics of trust. Even if the quotas are changed, the element of predictability in terms of what can be expected from transactions involving foreign governments will make many locals think twice about helping and make it more difficult for the soldiers who come to those countries without language skills.

Many miles from the conflicts in the Balkans, Iraq and Afghanistan I encountered another group of individuals who found themselves in a foreign country without language skills of their own. Their interactions with interpreters and translators, however, seemed to follow a familiar theme. In terms of access to public services, particularly medical and social services, researchers at University College London, UK reported:

> Our data suggest that the preconditions for communicative action are rarely met in the interpreted consultation. The interpreter's presence makes a dyadic interaction into a triad, adding considerable complexity to the social situation and generating operational and technical challenges. *Lack of trust*, intense pressure of time, mismatch of agendas (biomedical versus lifeworld), firm expectations of a specific outcome (e.g. referral, prescription) and profound power imbalances all promote strategic action (i.e. speech that seeks consciously or unconsciously to manipulate an outcome) rather than communicative action (i.e. sincere efforts to achieve understanding, and reach consensus) by all parties. In

consultations interpreted by family members (an option traditionally seen as 'second best' by policy makers), the social situation is very different. *Family members are generally trusted, share the lifeworld agenda, and shift the power balance in the patient's favour.* The interpreter occupies multiple social roles, including translator, interpersonal mediator, system mediator, educator, advocate, and link worker. The essence of professionalism in interpreting is shifting judiciously between these potentially conflicting roles. (Greenhalgh, Robb and Scrambler in *Social Science Medicine.* 2006 Sep; 63(5):1170-87. (Italics added)

In order to continue our discussion of trust in the language triad, let's assume that the first group we have discussed is a client who is paying for language services. The second element in the triad is made up of people who don't share a common language or culture with the client. We'll call them the client's *client.* They do share a language and perhaps cultural ties with the interpreter. The interpreter may find themselves in a maze of expectations and perceptions involving not only how each group sees them but how their relationship with the other group is perceived. At some point the observation of how well they speak the other language and understand the culture will lead to an unconscious question about where their loyalties lie. Can they be trusted as one of us or are they one of them?

In their essay *Using Interpreters to Access Services: User Views* (2004) Edwards, Alexander and Temple report minority users (your client's client) define a good interpreter as someone who "does more then just change words into another language. People prefer proactive interpreters who can empathise with them, help them understand procedures and plead their case."

If information exchange, as described by Brown and Yule, is polarized with one extreme represented by transactional discourse for the purpose of conveying "factual or propositional information" and the other extreme as interactional discourse, "language to establish and maintain social relationships." (Brown and Yule 1983:1-3); interpretation might seem to fall into some middle area. From the refugee-user perspective, Alexander, Edwards and Temple (2004, 4) reported, "Generally, however, the majority of people preferred drawing on their informal networks to act for them *in all but important legal and medical matters*. (Italics added.) The reason: "They trusted family and friends, rather than interpreters who were strangers to them; because they knew each other well and had a history of shared understandings and obligations." (ibid.) As an interpreter your trustworthiness may be judged by the perception of your obligation to the "other" language group.

In my book *Banking on Babel Survival & Profit Through Language* I relate several stories about the impact of interpreters who cross the line and coach or even change responses. In one example, an interpreter wants to spare a naive welfare applicant from the interpreter's home town the pain of having her children taken away and placed in foster care by a host government agency, so she coaches the applicant to give a response that will avoid that outcome. In another example a young person pressed into serving as an interpreter for peacekeeping forces in the Balkans changes the name of an ethnic group when she realizes the foreign military officer she is working with has misidentified the people he is talking to. My observations conclude with the comment, "Sometimes when the LSP (Language Service Provider) becomes an editor or coach, the first thing lost in translation is the truth. Other times you simply get to go home without getting your head bashed in or your throat cut." (Burgener, 2006:69)

Confidentiality tops the list of concerns expressed by people who need interpreters in order to get legal or medical services, but that aspect of trust comes with a particular price when a foreign language is involved. In a study of *Errors in Medical Interpretation and Their Potential Clinical Consequences in Paediatric Encounters* (Flores, et. al. 2003) for the Journal of Paediatrics, the choice of interpreters ranged from an 11 year old family member to nurses and social workers. The study reported "Errors committed by ad hoc interpreters were significantly more likely to be errors of potential clinical consequence than those committed by hospital interpreters (77% vs. 53%)." Dr. Flores and his team documented encounters involving children and parents with Limited English Proficiency (LEP) in a pediatric care department of a metropolitan hospital in the U.S.. The source of errors were identified as, "omission (52%), followed by false fluency (16%), substitution (13%), editorialization (10%), and addition (8%)." (Flores, et.al.2003, 8).

In Chapter 6, language in emergency medicine, Barbara Murphy, MD describes encounters from the perspective of an inner-city hospital in the United States and a military combat facility in the Balkans. Although the patients were quite different from Dr. Flores' the vulnerability created by language remained similar.

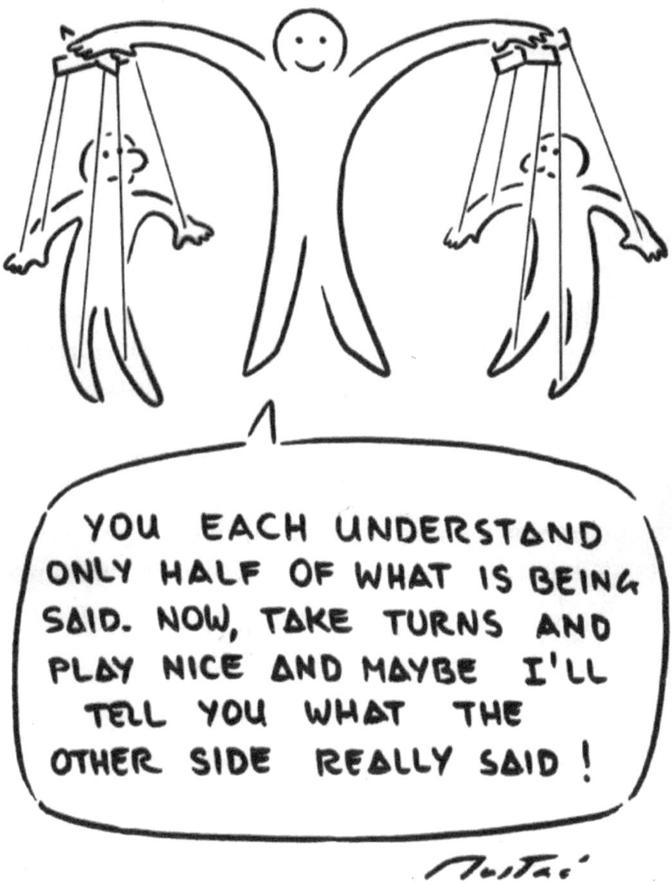

You Both Need Me and I Need Some Respect!

Trust is a very complex human behavior and as an interpreter looking for trust to be invested in you, the process may not always seem fair. Sometimes it's not. For those of us making the investment of trust, consider that our decision may reflect more on how we handle our own vulnerability than on anything you have said or done. Consider also the fact that there are people with no other access to a source of power beyond their ability to convince someone they can speak a needed language who complicate your life and ours. That's no excuse for behaving badly but is does serve to explain some of the skepticism you may encounter.

Edwards, Alexander and Temple offer the following insight from the work of Anthony Giddens (1984) on the concepts of 'personal' and 'abstract' trust.

In relationships that are based on personal trust, comprising bonds such as family or friends, people are assuming that they can rely on these familiar others' consideration of, and stable commitment to, their own needs, interests and preferences. Abstract trust is called upon when a person is not familiar with another party but is reliant on them for their expert knowledge and competence. It is based on Impersonal belief that a representative or member of a given expert group or institution will conventionally act according to particular principles, duties and requirements, which are *laid down in codes and training for professional groups*, rather than their own personal interests. (Italics added).

We all enter into relationships with professionals from various fields that we do not know personally but risk trusting based on our faith in the institutions which provide their credentials. The ability to begin a relationship with an interpreter based on knowledge of professional or government agency accreditation (*see ATA Chronicle Vol XXXV, 10:12 and XXXVI, I:31)* offers an external link to credibility. On the more important personal side, between the interpreter and English-speaking client, professional credentials create a perception of distance between the interpreter and what might be viewed (suspected) as too much shared culture. For the person with whom the interpreter shares a cultural or community connection, credentials which are properly presented before the session, can provide reassurance of confidentiality and independence from the government agency. (Edwards, et. al. 2004)

Trusting and being trusted are far more than abstract concepts when depending on someone in a foreign language environment. An appropriate response can save a life while one that is off the mark can cause more pain. Making the right choices is a role you have prepared for and accepted. Getting others who can only understand part of what you are saying to have confidence in your choices is another part of the risk and challenge. Start by trusting yourself.

Chapter 9

Internal Sources of Miscommunication

Six Points of Miscommunication

Professor Ralph K. White's book, *Nobody Wanted War: Misperception in Vietnam and other wars* (1968), contains valuable insights from the diplomatic message traffic that was exchanged between Austria and Serbia prior to World War I and between the government of Vietnam and several Western powers following World War II. Because his analysis was based on actual documents and background material from both sides, his observations and conclusions are more objective than political pundits who rely on conjecture. Perhaps the most useful information is a matrix developed by Professor White, a psychologist who taught at George Washington University, for describing the sub-conscious sources of miscommunication.

When exchanging information with each other in any language and particularly when working in a foreign language, through an interpreter, it becomes necessary to think beyond the words. How we see ourselves is one of the most complex emotional exercises we engage in as human beings. Like all forms of exercise, it is effective only when we do it on a regular basis. How we see others depends a great deal on how we see ourselves, however the process of understanding how we make decisions about other people offers opportunities to be objective without some of the personal, emotional baggage we carry into decisions about ourselves. An opportunity is just that: a chance to change course. Here are some of the route markers Dr. White offers to measure our progress—or at least a starting point.

The *diabolical enemy image* is a black and white approach in which we view others as always being bad. A *virile self image* causes us to view others as economically, intellectually or physically inferior. *Moral self image* casts the other guys as godless heathen, infidels, living—or worse yet—spreading their darkness. *Selective inattention* is a product of all the other elements of miscommunication and leads us to ignore what other's say because we are sure we have heard their lies before. *Absence of Empathy* helps us avoid awkward questions like what would I do in the same set of circumstances. Finally, *Military Overconfidence* sees others as stone-throwing mobs who would be no match for our technological superior firepower.

In my conversations with Dr. White, which began during my student days at George Washington University back in the 1970's and ended shortly before his death in 2007, I can tell you he was the first to acknowledge that there are diabolical people in this world who need to be contained and, when all other options fail, eliminated. Perhaps nobody wants war, but that doesn't mean it isn't necessary in some cases. There can be no empathy with people who engage in ethnic cleansing. The price of military overconfidence is paid in the lives of soldiers who must confront the reality of how ineffective a cruise missile is against a suicide bomber or Improvised Explosive Device (IED).

Language is our first line of non-lethal contact with other human beings. In the words of A Newspaperman's Prayer, news correspondent Bob Considine reminds us, " The 26 sharp-edged tools we call our alphabet can do what other tools do: build or destroy. Let me build. But let me know clearly, also, what should be destroyed, what darkness, what bigotry, what evil, what curse, what ignorance."

We've all been admonished to watch what we say. In working with an interpreter we must add, watch what others say for us.

Bibliography

Alexander,C. Edwards, R. and Temple, B. with Kanani, U. Liu, Z. and Miah, M. (2004) *Access to Services with Interpreters: User Views* Joseph Roundtree Foundation

Alexander,C. Edwards, R. and Temple, B. (2006) Contesting Cultural Communities: language, ethnicity and citizenship *Unpublished paper provided by the authors*

al-Khalidi, A., Hoffmann, S., Tanner, V. (2007) Iraqi Refugees in the Syrian Arab Republic: A Field-Based Snapshot *The Brookings Institution – University of Bern Project on Internal Displacement An Occasional Paper* Washington, DC.

Arocha, Izabel, President, International Medical Interpreters Association

Aspinall, P.J. (2007) Language ability: A neglected dimension in the profiling of populations and health service users *Health Education Journal, 66/1, 90-106*

Bancroft, M. (2003) 'Beyond the Language Barrier: What Managers Should Know about Laws That Govern Language Assistance for Immigrants in Publicly-Funded Programs and How to Meet These Service Delivery Challenges' *The Public Manager*, Vol. 32, 2003

Bancroft, M. (2005) The Interpreter's World Tour An Environmental Scan of Standards of Practice for Interpreters, *National Council on Interpreting in Health Care*

Barry, Ellen (2008) Barriers That Are Steep and Linguistic, *New York Times August 24,2008, The World section page 3*

Bischoff A, Bovier PA, Rrustemi I,Gariazzo F, Eytan A,Loutan L. (2003) Language barriers between nurses and asylum seekers: their impact on symptom reporting , and referral. *Soc Sci Med.* 2003 Aug;57(3):503-12 Department of Community Medicine, Geneva University Hospitals

Brown, G. & Yule, G. (eds) (1983) *Discourse Analysis* Cambridge Textbooks in Linguistics, Cambridge University Press, Cambridge, London, New York

Budd, S. & Rusbridger, R. (eds) 2005, *Introducing Psychoanalysis Essential Themes and Topics,* Routledge, London

Burgener, R. (2005) *Banking on Babel Survival & Profit Through Language* Publish America, Baltimore

Caper, R. (1988) *Immaterial Facts: Freud's Discovery of Psychic Reality and Klein's Development of his Work,* Jason *A*ronson, Inc. Northvale, NJ, London

Drenen, G, (1996) Counting the cost of language services in psychiatry.__*South African Medical Journal.* 1996 Apr;86(4):343-5 Department of Psychiatry, University of Cape Town.

Edgcumbe, R. (1981) 'Toward a Developmental Line for the Acquisition of Language' *Psychoanalytic Study of the Child* 36;71-103

Edwards, R., Alexander, C. and Temple, B. 'Interpreting Trust: Abstract and Personal Trust For People Who Need Interpreters to Access Services' *Sociological Research Onlin*e Vol.11, Issue 1

Encarta World English Dictionary (1999) St. Martin's Press, New York

Erikson, Erik (1985) Pseudospeciation in the Nuclear Age *Political Psychology Vol. 6 No.2 pp.213-217*

Flores G, Abreu M, Schwartz I, Hill M (1999) The importance of language and culture in pediatric care: case studies from the Latino community. *PEDIATRICS Vol. 103 No. 6 June 1999, pp. 1253-1256*

Flores, G., MD, (1) Laws, M. B. PhD, (2) Mayo, S. J. EdM, (2) Zuckerman, B., MD, (3) Abreu, M. MD, (1,3) Medina, L., MD (3) and Hardt, E.,MD (4) (1) Center for the Advancement of Urban Children, Department of Pediatrics, Medical College of Wisconsin, Milwaukee, Wisconsin, (2) Latino Health Institute, Boston, Massachusetts (3) Departments of Pediatrics (4) Internal Medicine, Boston University School of Medicine, Boston, Massachusetts, (2003) 'Errors in Medical Interpretation and Their Potential Clinical Consequences in Pediatric Encounters' *PEDIATRICS Vol. 111 No. 1 January 2003: 6-14*

Flores, G. MD, (2005) 'The Impact of Medical Interpreter Services on the Quality of Health Care: A Systematic Review' Medical College of Wisconsin *Medical Care Research and Review*, Vol. 62:3, 255-299

Framer, I. (2003) Chair, National Association of Judiciary Interpreters and Translators *Summit Lorain Study* http://languageaccess.us

Freud, S. (1937) *Constructions in Analysis*, Hogarth Press, London

Henrard, K. (2000) International Journal on Minority and Group Rights vol 7, no.2 Feb 2000

Hinshelwood, R.(1991) *Dictionary of Kleinian Thought* Jason Aronson, Inc. Northvale,NJ, London

Hunt, J., (1983) *Psychoanalytic Aspects of Fieldwork* Sage Publications, Newbury Park, London, New Delhi

Hunt, L. and de Voogd, K. (2007) 'Are Good Intentions Good Enough?: Informed Consent Without Trained Interpreters' *Society of General Internal Medicine*, 2007:22; 598-605.

Johansen, B.E. (1990) 'Native American Societies and the Evolution of Democracy in America, 1600-1800' *Ethnohistory*, Vol. 37, No. 3 (Summer, 1990), pp. 279-290

Jones, D. (2007) 'Should the NHS curb spending on translation services? *British Medical Journal (BMJ) 24 February 2007;334-399*

Kant. I.,(1795), *Perpetual Peace: A Philosophical Sketch* Available at: file://M:\Immanuel Kant Perpetual Peace.htm

Karlsen W.B, Haabeth A.L. (1998) 'The telephone interpreter. A good alternative to the traditional interpreter *Tidsskr Nor Laegeforen. 1998 Jan 20;118(2):253-4*'

Kelly, L.G.,1979, *The True Interpreter, A History of Translation Theory and Practice in the West,* Blackwell, Oxford,

Klein, M. (1928) 'Early Stages of the Oedipus Conflict' *International Journal of Psychoanalysis* in *Love, Guilt and Reparation & Other Works 1921-1945;188-198,* Delacorte Press/Seymour Lawrence 1975

Klein, M.(1946) 'Notes on some Schizoid Mechanisms' *International Journal of Psychoanalysis* and in Dictionary of Kleinian Thought, Hinshelwood, R. 1991,79

Klein, M., Heiman, P. and Money-Kyrle, R. (1955) *New Directions in Psychoanalysis*, Tavistock, London

Labov, W., (1972) *Language In The Inner City Studies in the Black English Vernacular*, University of Pennsylvania Press,

Loentz, E., (2006) Yiddish, Kanak Sprak, Klezmer, and HipHop: Ethnolect, Minority Culture, Multiculturalism, and Stereotype in Germany, *Shofar: An Inter-disciplinary Journal of Jewish Studies* -Volume 25, Number 1, Fall 2006: 33-62

Mikkelson,H. (2002) *Introduction to Court Interpreting* St. Jerome Publishers Ltd.Manchester, UK

Nimni, E. (2007) Constitutional or Agonistic Patriotism? The dilemmas of liberal nation states. To be published in *Constituting Communities: Political Solutions to Cultural Differences* Mouritsen,P. Jorgensen,K.E. (eds) Polgrave

Paz Beltran Avery, M. Ph.D (2001) The Role of the Health Care Interpreter An evolving dialogue *The National Council on Interpreting in Health Care Working Papers Series* at www.nchic.org

Pegelow, T., (2006), *Determining 'People of German Blood', 'Jews' and 'Mischlinge': The Reich Kinship Office and the Competing Discourses and Powers of Nazism*, 1941–1943. Department of History, Grinnell College Contemporary European History *(2006), 15: 43-65* Cambridge University Press

Petevi, M. (ed) (2001) Declaration of Cooperation Mental Health of Refugees, Displaced and Other Populations Affected by Conflict and Post-Conflict Situations, *World Health Organization,* Geneva WHO/MNH/MHP/99.3 Rev.1.

Rudmin, F. W. (2003) Critical history of the acculturation psychology of assimilation, separation, integration, and marginalization *Review of general psychology* vol. 7, n°1, pp. 3-37

Rzucidlo, S.F. (2003) Premise Alert System *www.premisealert.com*

Spalding, M. and Ortega, I. (2007) 'Immigration reform: the need for upholding our national language' *http://www.heritage.org/Press/Commentary/ed080207e.cfm.*

Temple, B. (2002) 'Crossed Wires: Interpreters, Translators, and Bilingual Workers in Cross-Language Research' *Qualitative Health Research,* Vol.12, No. 6 July 2002

Temple, B. (2005) 'Nice and Tidy: Translation and Representation', http:/www,socresonline.org.uk/10/2/temple.html

Temple, B. (2006) 'Being Bilingual: Issues for Cross-Language Research' *Journal of Research Practice* Vol, 2, Issue 1. Article M2

Van Oudenhoven, J.P., Prins, K. S., Buunk,B. P. (1998) 'Attitudes of minority and majority members towards adaptation of immigrants' *European Journal of Social Psychology* Volume 28, Issue 6, 995 - 1013 Author(s) Affiliation: University of Groningen, The Netherlands

Volkan, Vamik, M.D.,DLFAPA,FACPsa,(2006) *Killing in the Name of Identity*, Pitchstone Publishing, Charlottesville, Virginia

White, Ralph K. (1968) *Nobody wanted war: Misperception in Vietnam and other wars Garden City, N.Y: Doubleday.*

Wilson-Stronks, A.,Lee,KK, Cordero,CL,Kopp,AL,Galvez,E. (2008) *One Size Does Not Fit All: Meeting the Health Care Needs of Diverse Populations* Oak Brook Terrace, IL. The Joint Commission

About the Authors

Robert D. Burgener, BA, MA.

Robert D. Burgener served as an enlisted photo-journalist with *Pacific Stars & Stripes* during Vietnam (1968-1975) and later with ABC Television News and Associated Press Radio. He rejoined the U.S. Army Reserve in 1988 completing Warrant Officer Candidate School as a military intelligence analyst and interrogator. He served as a Force Protection Officer with Task Force Eagle, Bosnia in 1999. His last deployment in 2004, 2005 prior to retirement was as an interrogation screening officer at Abu Ghraib, Iraq and analyst with the Iraq Survey Group. Between military assignments he founded International Connections (INTERNECT) providing direct support to refugees and displaced persons through various organizations in the Former Soviet Union. Today INTERNECT provides communication risk analysis to international organizations.

He studied at the University of Vienna, Austria and the University of Madrid, Spain before completing his BA in General Studies at George Washington University. After a refresher course at the Jung Institute in Zurich, Switzerland he finished a

MA in Refugee Care at the Center for Psychoanalytic Studies at Essex University and London's Tavistock Clinic in the UK. Robert and his wife Irina have three children.

Barbara J. Murphy, MD.

Dr. Murphy is an emergency physician certified by the American Board of Emergency Medicine. She has practiced medicine for over thirty years at hospitals in the city of Philadelphia and currently serves on a regional emergency management committee. A former Colonel in the U.S. Army Reserves, she commanded a combat support hospital and was deployed with Operation Joint Endeavor and Operation Joint Guardian. She commanded the medical facility at Camp Able Sentry in the Former Yugoslav Republic of Macedonia during the summer and fall of 2001. On September 11, 2001 news of the terrorist attack on the United States held particular significance because her daughter Monica, a Sergeant First Class in the U.S. Army Reserve was serving at the Pentagon. Monica survived.

Dr. Murphy received her B.S.N and M.A. from the University of Iowa and her M.D. from the Medical College of Pennsylvania.

Barbara has seven children, 17 grandchildren and three great-grandchildren.

INDEX

A

B

D

E

F

M

N

P

T

U

V

W

Z

www.ingramcontent.com/pod-product-compliance
Lightning Source LLC
LaVergne TN
LVHW101209200926
848479LV00018B/221/J